SCARD, Ann
The building stones of Shropshire.

The Building Stones of Shropshire

M. Ann Scard

First published in the U.K. in 1990
by Swan Hill Press, an imprint
of Airlife Publishing Ltd.

British Library Cataloguing in Publication Data

Scard, M. A.
The building stones of Shropshire.
1. Shropshire. Stone buildings
I. Title
721'.0441

ISBN 1 85310 066 8

Swan Hill Press

An Imprint of Airlife Publishing

101 Longden Road, Shrewsbury SY3 9EB, England.

Contents

Introduction

1: Sandstones 1

2: Grinshill Stone 9

3: The New Red Sandstone of North Shropshire 29

4: The New Red Sandstone of East Shropshire 38

5: The Old Red Sandstone of South Shropshire 49

6: Carboniferous Sandstones 58

7: Sandstones of the Ancient Seas 67

8: Limestones for Building Use 73

9: Miscellaneous Building Stones 78

Glossary 89

Index of Buildings 96

Bibliography 99

INTRODUCTION

Shropshire has few rivals amongst the other counties of Britain for variety of landscape and architecture. This is largely due to its impressive array of geological outcrops which not only contributes to the variable relief and soils, with their associated patterns of natural vegetation, but also produces a diversity of potential building materials including stone blocks quarried from solid rock faces and a variety of unconsolidated gravels, sands and clays used for the manufacture of aggregate and bricks. Geological factors have also influenced the development of differing agricultural systems and industrial enterprises with their various demands for specific types of buildings. Prior to the mass production and increased mobility of building materials, communites were very dependent on locally available supplies and therefore a study of buildings closely reflects the distribution of the underlying rocks. Geology is also implicated in the actual practicalities of construction since the foundations of buildings are affected by the natural arrangement and drainage of the rocks beneath. Close links can therefore be recognized between the geological endowment of Shropshire and the types of buildings found within the county.

Shropshire has a well-balanced distribution of lowlands and uplands. The lowlands are concentrated in the north and east and border the counties of Cheshire and Staffordshire. The uplands dominate the west and south and merge with the Welsh foothills and the undulating landscapes of Hereford and Worcester (Fig. 1). For the most part, the younger Carboniferous, Permo-Triassic and Triassic rocks prevail in the low-lying districts, and a wide range of older rocks dominates the uplands (Fig. 2 and Fig. 3). The older rocks give rise to the most spectacular and diverse scenery of the county including the chain of steep-sided 'hogs-backs' (Fig. 8), such as Caer Caradoc, which is located to the east of the Church Stretton valley, and also the more subdued but equally arresting plateau landscape of the Longmynd to the west. Both types of landscape evolved on ancient Precambrian outcrops. Elsewhere in the upland districts a variety of Cambrian, Ordovician, Silurian and Old Red Sandstone rocks has been differentially eroded into a series of hills and vales, low escarpments such as Wenlock Edge, having formed wherever the dip of the rocks has encouraged the evolution of opposing steep and gentle slopes (Fig. 8).

Igneous rocks of different geological ages are often found in proximity to hill summits for they tend to be tough, compacted rocks resistant to weathering. The comparatively young Carboniferous dolerite, known locally as 'dhustone', caps the Clee Hills; much older Precambrian igneous and metamorphic rocks outcrop on Wart Hill, Pontesford Hill and the Wrekin in addition to those hills east of the Church Stretton valley previously referred to. Such rocks constitute tough building stones best suited to rubblestone masonry. They break down into excellent roadstones.

Pleistocene ice which spread southwards from the highlands of Scotland and Cumbria and eastwards from the mountains of Wales during the Great Ice Age carried vast quantities of detritus which later mantled much of Shropshire when the melting ice jettisoned its load. Gravels, sands and clays, though unconsolidated, are classed as rocks and as such augment the geological resources of Shropshire. They are concentrated in the north of the county especially in the Ellesmere district where the meres and mosses proclaim the whereabouts of former stagnating ice. However, streams draining east and south have re-worked and transported the loose material so that it may now be extracted downstream, for example, in the valley of the River Severn near Bridgnorth.

This brief review of the geology and related landscapes of Shropshire serves to suggest the very wide variety of stones available for building purposes, as rocks of Precambrian to Jurassic age are each represented in the county to some degree. Within this range, certain stones are potentially far better building material than others and have usually been exploited commercially for distant markets. Others have only been of use in their immediate locations of outcrop, often because alternative materials were in short supply.

The following chapters have been arranged according to building stone type, for example, sandstones, limestones etc., because the inherent characteristics of stones as building material depend more on their intrinsic physical and chemical properties than on their relative ages. Reference is only made to chronological order whenever it provides clues to the potential of stones as building material.

For those wishing to consult systematic studies of the geology of the county there are numerous published sources available, for this is an area that has attracted the attention of leading geologists for decades. Relevant books and articles are listed along with the appropriate maps in the bibliography, in addition to more general books on landscape history and architecture. This study explores the links between geological resources for construction purposes and the diverse buildings required during successive phases of economic and social development, and merely trespasses on the established preserves of the geologist and landscape historian without making any attempt at a definitive statement in either discipline.

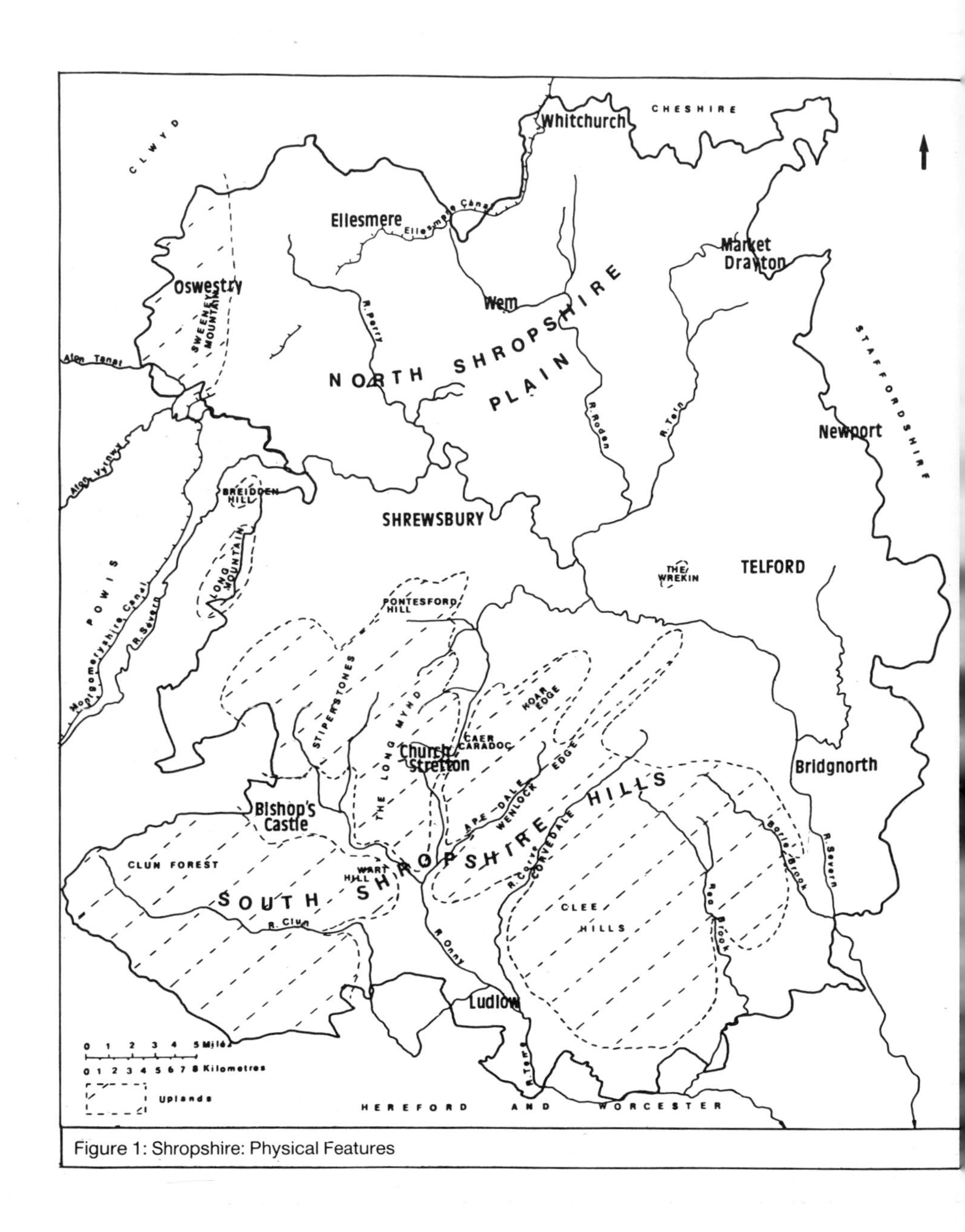

Figure 1: Shropshire: Physical Features

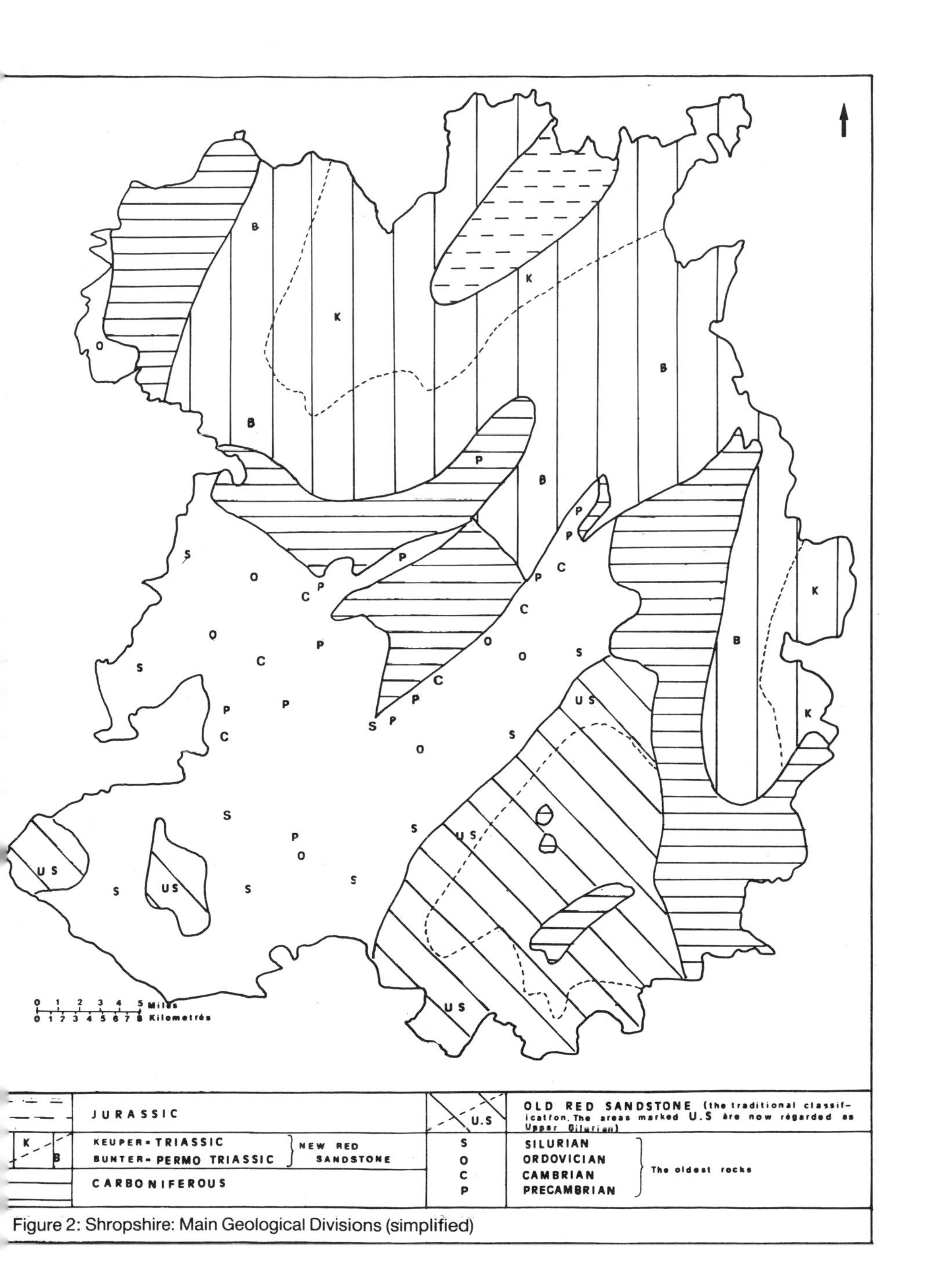

Figure 2: Shropshire: Main Geological Divisions (simplified)

Figure 3: Shropshire Stones and the Geological Column

(Only the stones used in construction are included)

Millions of years	Period	Stones
	RECENT PLEISTOCENE	Mostly sand and gravel.
50–150	Younger 'solid' rocks not found in Shropshire	
150	JURASSIC	Wem – Prees area. Some local building limestones.
200	TRIASSIC	Keuper Beds produce best building freestones. Nesscliffe, Grinshill, Harmer Hill. Myddle, Hawkstone, etc. Bunter Beds (now designated Permo-Triassic) include Mottled Sandstones and Pebble Beds — local building stones in Bridgnorth area especially.
250	PERMIAN	Alberbury Breccia — Cardeston district west of Shrewsbury — for local buildings.
300	CARBON-IFEROUS	Good building freestones amongst the Coal Measures, especially in the East Shropshire coalfield and south of Bridgnorth – Highley, Alveley and Arley. Good building freestones in the Millstone Grit formations — Sweeney Mountain and the Clee Hills. Tough limestones for roadstones and lime but some building stones obtained from the Carboniferous Limestone in the Oswestry district, the East Shropshire coalfield and the Clee Hills. Some polished for 'marble'.
350	OLD RED SANDSTONE (DEVONIAN)	Farlow Sandstone — Clee Hills. Various sandstones from the Ditton Series and the Clee Group for local buildings. Best buildings stones are the Holdgate and Downton Castle Sandstones found in Corvedale, Clun, Onibury districts (on Geological Survey maps in the Old Red Sandstone areas).
400	SILURIAN	(Holdgate and Downton Castle Sandstones now classed as Upper Silurian rocks) Aymestry and Wenlock Limestones — used for roadstone, lime mortar and as building stone for local use on vernacular buildings. Ludlow Shales — used for building but weather very badly. Kenley Grit — used for local building in the Kenley, Harley district.
450–500	ORDOVICIAN	Acton Scott Limestone — for local building. Cheney Longville Flags — for roofing slates. Horderley/Soudley/Chatwall Sandstone — best building sandstone for local churches, farms, cottages. Hoar Edge Grit — a tough sandstone in Onny valley and Hoar Edge districts — used for local buildings. Includes flaggy beds for stone 'slates'. Stiperstones Quartzite — very tough rock for roadstone and rubblestone — Pontesbury, Nills Hill. The igneous rocks found in the Shelve district etc. mostly for roadstone; in the Breidden Hills used for roadstone and for feldspar for glazing china.
	CAMBRIAN	Comley sandstone of Church Stretton district used for local buidings. Wrekin Quarzite — the Wrekin and the Ercall for roadstone.
550	PRECAMBRIAN Not to scale	Conglomerates and grits used for tough building stone and roadstone — eg. Bayston Hill. Igneous rocks at the Wrekin, the Ercall and Earl's Hill etc. used for roadstone and rough walling. Vernacular buildings of the Longmynd district are constructed of assortments of Precambrian rocks.

1
Sandstones

Sandstones constitute the largest category of building stones in Shropshire. They are prevalent in the Triassic and Permo-Triassic areas of the North Shropshire Plain, eastern Shropshire and the Severn valley district to the south-east. Most of the successful commercial quarries of the county developed here. Other quarries were opened on the Carboniferous sandstones associated with the coalfield districts of eastern Shropshire, the Severn valley, the Clee Hills, and the Oswestry and Shrewsbury districts. Elsewhere in the county geologically older sandstones were generally restricted to local use. These included sandstone beds within the Old Red Sandstone outcrops of the Clee Hills, and Corvedale and the Ordovician and Silurian outcrops of the Ludlow, Craven Arms, and the Scarp and Vale districts of south Shropshire. In the Clun Forest Basin and the Long Mountain areas Silurian sandstones were of limited local use since they usually provided inferior quality building stones. Shropshire's most productive and profitable quarries were concentrated in the north and east of the county where the sandstones yielded the best freestones. For centuries these proved to be the most useful and attractive building stones and their appeal widened in the nineteenth century with the growth of urban centres in the Northwest and the Midlands and with the provision of rail transport to these expanding markets. Shropshire sandstones competed successfully with comparable stones from Yorkshire, Staffordshire, Derbyshire and Nottinghamshire.

The nature of sandstones varies according to their constituent minerals and the way these are bonded to form rock. Quartz is the common element. It registers seven on Moh's scale of mineral hardness with diamond, the ultimate mineral in this respect, registering ten. As a result, quartz is extremely useful for cutting and grinding purposes and has been used for centuries for slicing into stone slabs by the simple expedient of applying quartz grains to the cutting and sawing edges of hand- and machine-operated tools. Alveley sandstone provided grindstones for the Stourbridge manufacturers of agricultural machinery until the early part of this century.

Quartz, in addition to being hard, is chemically stable because it is composed of silica. It is therefore resistant to any decay induced by atmospheric pollutants. The twin attributes of physical hardness and chemical inertia ensure a good survival potential for quartz grains in different natural environments, some of which can be extremely punishing. On coastal beaches they withstand the battering of waves and in deserts the mutual bombardment of flying sand grains carried at high wind velocity. With prolonged attrition, however, they reduce to fine dust.

It would seem reasonable to expect sandstones to provide sound, durable building material, and this is often the case. The productive Grinshill, Highley and Alveley quarries of Shropshire have each in the past supplied quality building stones. Highley stone was considered as a possible choice for the construction of the proposed new Houses of Parliament in 1839. All three types of stone have featured on the facework of major bridges across the Severn between Shrewsbury and Gloucester — a severe test of any stone because of the recurring menace of flood and impact damage. Grinshill stone features prominently in the fabric of several prestigious Shropshire buildings of various dates, functions and architectural styles and also on a number of municipal buildings in other counties.

Not all the sandstones of Shropshire are so reliable. Even in proximity to the commercial quarries referred to, stones of comparable geological origin were extracted for local buildings, some of which are now decidedly the worse for wear. This is especially true of vernacular premises in the Permo-Triassic districts of the North Shropshire Plain, eastern Shropshire around Newport and in the Severn valley of the Bridgnorth area. Here, any churches constructed of the bright red sandstones continually require repairs to their masonry fabric, even though several were built in the latter half of the nineteenth century when the advancing sciences of Chemistry and Geology were shedding more light on the wearing potential of building stones. Not all local architects and masons were persuaded by their findings and continued to use the less reliable stones of the immediate locality.

Several factors explain the unreliable nature of some sandstones despite the inclusion of quartz. Even though they are consistently tough and chemically stable, individual quartz grains may differ in physical form. Those that have been subjected to desert conditions tend to be rounded due to the continual impact of flying grains. Many of the Permo-Triassic outcrops of Shropshire consist of millions of these rounded grains which even when densely packed can only touch at relatively few points of contact on adjacent spheres. The resulting rock is potentially very friable when the binding substance in the spaces between is chemically unstable. Voids permit the seepage of mineral-laden moisture which is very damaging to the stone. Many of the red sandstone outcrops of north and east Shropshire formed under arid and semi-arid conditions so that the rounded quartz grains are cemented by a coating of iron oxide. Such ferruginous sandstones are notoriously friable; even a cursory rubbing of the stone causes the shedding of numerous grains. Such material is often put to better construction use as building sand rather than building stone.

Angular quartz grains have the potential to form a more densely compacted rock mass since the individual grains can interlock to produce a tighter fit. These angular grains, weathered out of parent rock such as granite, must have been protected by subsequent transportation in water rather than by wind. Suspended in reasonably quiet waters, the grains can maintain their sharp edges. Given prolonged and relatively undisturbed sedimentation these jagged grains often accumulate as gritty sandstones and, providing the binding matrix is chemically stable, they usually constitute the toughest of building stones, most of which were exploited by the railway and waterworks engineers of the nineteenth century. The Carboniferous sandstones, including the very tough Millstone Grit, come into this category. They were used as building and engineering stones in the Oswestry-Morda, East Shropshire and Severn valley coalfield districts of Shropshire.

The sandstones that appealed most to the architect and mason were usually finer grained than the grits, which could sometimes be too tough and gritty to work to a smooth and refined finish, and were moreover a major cause of lung damage and injury to the men working in the stone trades. Fine and evenly grained sandstones, especially those cemented with silica, were the most attractive since they were relatively easy to dress, mould and carve, but at the same time had the inherent resistance to weathering essential for the preservation of carved details. Grinshill stone is Shropshire's prime example of a quality sandstone, having a silica content of 95.46 per cent.

The ferruginous and siliceous sandstones, briefly referred to above, serve to illustrate the widely ranging merits of sandstones as building material. Other sandstones are classed as argillaceous if they are clayey — not the most promising of building stones — micaceous if they contain a high proportion of shiny mica flakes, and calcareous if the calcium carbonate content is high. The latter are often as vulnerable to atmospheric pollutants as many limestones.

Siliceous sandstones such as Grinshill stone, in addition to providing some of the most reliable building stones, are extremely useful for engineering purposes. Waterworks engineers involved in the construction of dams in the mountain regions of Britain, where the waters are inclined to be acidic, require a chemically resistant stone. Grinshill stone was used in the construction of the Elan Valley Waterworks for Birmingham Corporation at the turn of the century. The high silica content also afforded this stone some measure of protection against the ravages of salt-water spray and for this reason it was used for several prestigious buildings erected in the Welsh coastal resorts during the last century. No stone, however, is entirely immune to chemical attack and, in any case, there is a vast array of physical disintegration that results from other hazards including temperature changes, in particular, frost damage. It is only a matter of time before stone of any calibre decays as it succumbs to the natural processes involved in the recycling of the materials that constitute the fabric of the earth's crust. What is crucial to the survival of stone buildings is the relative rate of that decay.

Stones of architectural appeal obviously need to be a pleasing colour. Colour varies according to the chemical content of the stones, for example, the presence of iron is responsible for the red tints found in so many of the Permo-Triassic rocks. The mineral glauconite produces green tints; limonite yellow. A high proportion of silica usually produces a pale-coloured sandstone. Streaks of different colours are sometimes evident in a rock and, although attractive in some circumstances, this variegated material is never suitable for formal styles of classical architecture which depend for their chaste and dignified effect on a reliable supply of uniformly pale-coloured stones. Castellated styles can tolerate the bright red sandstones of north and east Shropshire.

Although colour is important, the textural uniformity of stones, when combined with the dimensions at which they are available at the quarries, must rate of equal importance and in some cases be the determining factor in selection. Some sandstones vary too much in texture to be significant as building stones except for the crudest of structures such as farm outbuildings and walls. Such sandstones include those with clay partings and clay galls which, apart from constituting blemishes, are potential sources of weakness, becoming conspicuous as grooves and cavities as the stone weathers. These argillaceous sandstones are very unreliable.

Conglomeritic sandstones have the most pronounced variations of texture since they are composed of many pebbles of different sizes embedded in a sandstone matrix. These stones are so lumpy that they are virtually impossible to cut and dress. Conglomerates are essentially the hardened masses of ill-sorted sediments which accumulated in river beds and on the shores of seas and lakes in former times. They are common amongst the Permo-Triassic sandstones of Shropshire, especially those of the Bunter Pebble Beds, for very good reason.

In the present day climate of the British Isles the rivers are fed by fairly constant precipitation so that the loads of sediment they transport are gradually sorted out along their respective routes to the sea, the finer detritus being sifted out and carried along for greater distances downstream. In contrast, desert streams are intermittently fed by infrequent but sometimes torrential rainstorms so that rapid flows carry and deposit ill-sorted loads of clays, sands, pebbles and boulders. As there is little vegetation to check the swift and destructive torrents, much erosion takes place in a short time but the floodwaters subside equally quickly, sinking away into the parched earth and rapidly evaporating in the hot sun. As there is insufficient time for sorting and grading debris, the boulders, sands and clays are dumped indiscriminately to form the conglomerates of the future. Such desert conditions prevailed in Permo-Triassic and Triassic times leaving the legacy of conglomeritic sandstones now found in Shropshire and adjàcent parts of Cheshire and Staffordshire. Some of these are far too coarse to be of any use for building purposes but those made up of relatively few pebbles and a cohesive sandstone matrix can make good building stones, but only rarely and never successfully for ashlared masonry.

In terms of potential durability, colour and texture, many rock outcrops would appear to be promising building material, but there is one other factor of major importance which also depends on the mode of formation of the

Sections through stratified rocks

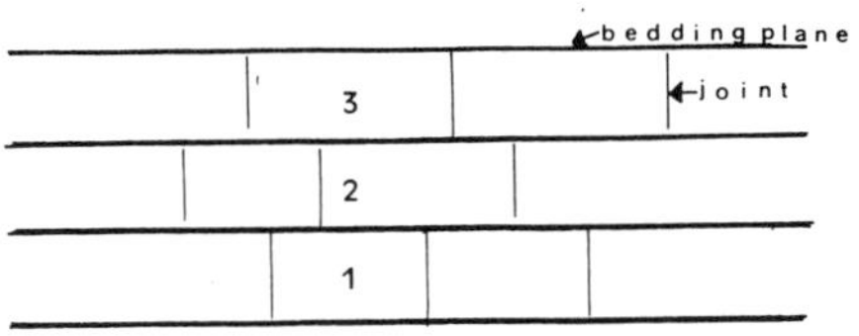

bedding plane

joint

3

2

1

1 Regular Bedding

1→2→3 progressively younger beds

2 Graded Bedding

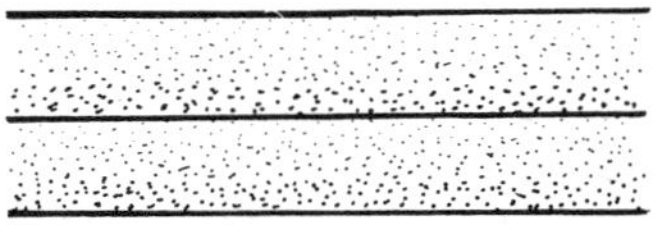

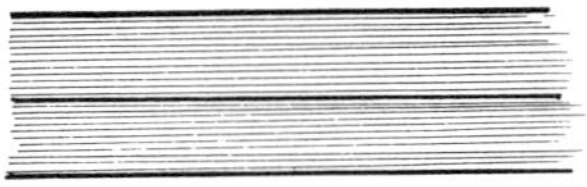

3 Laminated beds

Sections through cross-stratified (false-bedded) rocks

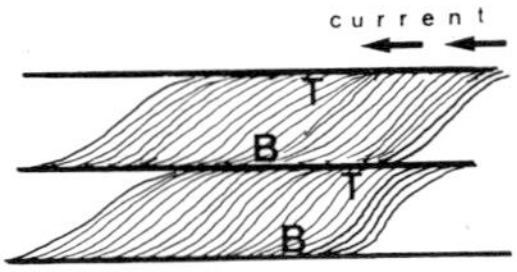

current

T

B

T

B

1 Current Bedding

T Top-set beds
B Bottom-set beds } of current bedding units

current

B

B

2 Current Bedding (with truncated top-sets)

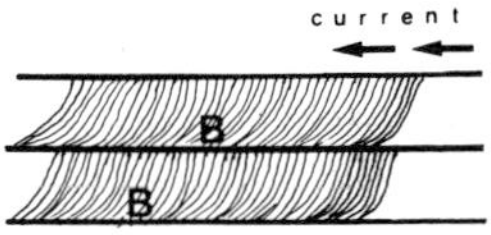

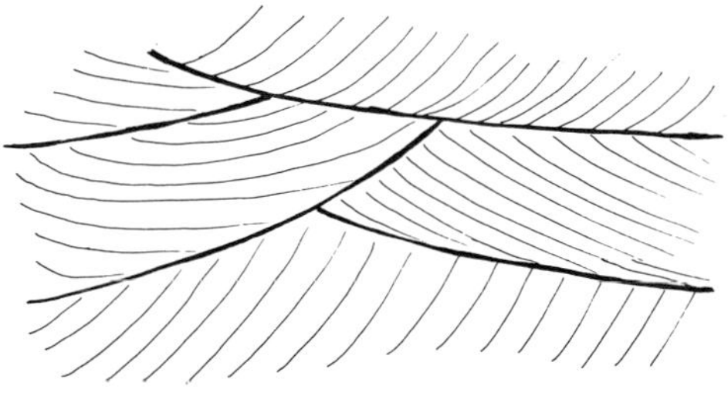

3 Dune Bedding

(Complex patterns develop with frequent changes of wind direction)

Figure 4: Types of Bedding

rocks. Quarries yielding a steady output of large dimension blocks of quality stones are comparatively rare, simply because most modes of deposition preclude the formation of very deep beds of uniform and horizontally disposed material which contains relatively few vertical partings known as joints (Fig. 4).

The horizontal partings which divide beds of rock mark pauses in the laying down of sediments such as clays and sands; the vertical cracks — the joints — formed as the sediment dried out, hardened and contracted. Quarrymen exploited these natural divisions when extracting blocks from the quarry face, preferring horizontal or slightly dipping beds to those with steeper and more difficult to manage angles. Not all beds of rock are conveniently arranged and of sufficient depth to provide the kind of massive blocks essential for the construction of bridges and the lower storeys of mansions and major public buildings. Quarries supplying such dimension stones of freestone quality were keenly sought by leading architects, bridge engineers and masons.

Reasonably quiet conditions of sedimentation are necessary for the accumulation of deep layers of uniform material which slowly settles down with consistently well-sorted grains. Deep, still waters and relatively shallow and undisturbed waters located over gently subsiding basins create the perfect environments for such massive accumulations of rock material. Even then, there can be no guarantee of an endowment of sound building stones for use in the distant future. Several episodes of earth movement throughout geological time have caused major dislocations of the earth's crustal rocks so that the incidence of favourably disposed rocks for quarrying purposes is not as great as might be expected.

Fast-flowing river currents, agitated tidal waters and wind-blown locations are not likely to produce the thick beds of rock necessary for monumental architecture. Many outcrops are what is termed 'false-bedded' or 'cross-stratified', that is, the layers of rock are disposed in sloping sequences, often tilted at different angles. Current-bedding is found wherever swift-flowing streams repeatedly changed course and shifted sediment around before it was finally permitted to settle. Dune-bedding results from the piling up in dune formation of wind-driven sand and its subsequent re-arrangement by changes of wind direction so that complex patterns of cross-stratification developed. Deltas which built out from the shores of former lakes and seas also formed sloping layers of sediment which can now be recognized in certain rock formations. Such variants of cross-stratification are common amongst sandstones of all geological ages but are especially conspicuous amongst the Permo-Triassic outcrops of Shropshire. The Bridgnorth district, for example, is renowned for its dune formations. Current-bedded sandstones of Carboniferous age are especially common amongst the Millstone Grit formations since these accumulated on deltas prior to the establishment of the Coal Measure forests. Cross-stratification detracts from the appeal of some sandstones for building purposes for the stripey effect can be visually disturbing if too obvious. Blocks of stone destined for exposed sites on the corners and projections of buildings should never be selected from cross-stratified layers since these are prone to differential weathering and often develop a fluted profile. Some cross-stratification, however, is not too pervasive and may actually enhance the colour and textural variations within individual stones so that the resulting masonry lacks the bland appearance of perfectly uniform stone.

Although thinly bedded sandstone outcrops are incapable of yielding the big dimension blocks essential for the heavy, dignified walling masonry of imposing buildings, they have often been quarried for modest vernacular premises, since they are ideal for rubblestone walling which is often stout and durable and visually in accord with a rural setting.

Some of the most thinly bedded sandstones are micaceous, that is they contain a high proportion of flat mica flakes which often concentrate in layers between the heavier sand grains so that the rock splits into thin slabs ideal for the supply of hearthstones, flagstones and roofing material. These stone 'slates' should not be confused with the famous slates of North Wales which are true metamorphic slates derived from ancient shales and clays which were subsequently compressed and hardened by the major earth movements responsible for the creation of the mountain ranges. Shropshire's stone 'slates' are entirely sedimentary in origin and essentially the product of the slow differential settling out in water of alternating shallow layers of quartz and mica. The most famous of these are the Harnage 'slates' of the Hoar Edge escarpment which were used for the roofing of Pitchford Hall, the Prior's Lodge at Much Wenlock and also Madeley Court.

Although sandstones vary so much in terms of their durability, colour, texture and the sizes of blocks available, most have the crushing strength to withstand the pressures brought to bear on load-bearing walls. Tests on samples of Grinshill stone at the end of the nineteenth century showed its crushing strength to be in the order of some 171.5 to 245.9 tons per square foot and well above the limits required for large buildings. St. Paul's Cathedral, for example, was calculated to have a maximum stress of fourteen tons per square foot. Even in bridges, the compression on the main keystone of the central arch is well below the crushing strength of most sandstones, though it is obviously prudent to use tough varieties such as Grinshill and Highley stones for major bridges.

Before discussing Shropshire's sandstones in more detail, it is necessary to emphasize two factors of great significance. The first concerns the vital matter of the seasoning of freshly quarried stone. Sir Christopher Wren is alleged to have rejected any stones for the construction of St. Paul's which had not been seasoned for at least three years. Seasoning entails exposure to the open air of any recently quarried 'green' stone which still contains its quarry sap (the mineral-laden moisture present in stone embedded in solid bedrock). Most stones are 'soft' on their initial detachment from the rock body and only harden as they dry out in the atmosphere. The seasoning of stone is an essential preliminary stage to the cutting, dressing and final fixing of masonry components in order to maximize the wearing potential of stone in a building. Stone sealed in the bedrock for millions of years can hardly be expected to withstand the ravages of wind, rain and temperature changes — in particular frost — in the purely artificial environments of man-made structures especially if, as is often the case, these are located in towns with severe pollution problems.

The second factor concerns the identification of the natural bed of the stone. The original sediment accumulated in layers which are readily recognizable in some stones but not so easily in others. It is most important that the mason should fix a stone parallel to the natural bedding which, unless the sediment was cross-stratified, conforms to the steady building up of the layers of sediment in horizontal sequences. In this way he can enhance the stone's resistance to weathering and minimize the effects of compression. In bridge construction the arch stones need to be laid with the bedding placed downwards around the radius of the arch but in buildings most of the compression is vertical and the bedding of each block should be horizontal. Stones laid at odds with the natural bedding are liable to spall or peel on the face and, more seriously, to fracture and fail as masonry components. It is sound quarry practice to mark the natural bedding as the blocks are detached from the bedrock especially if the stratification (layering) is inconspicuous.

The selection of general points raised in this chapter is intended to provide some background information to the more detailed descriptions of the specific sandstones quarried in Shropshire. Other factors of importance will emerge as these various stones are discussed for it is virtually impossible to make any valid generalizations about the characteristics of stones. No two quarries on the same geological outcrop produce identical material and even within a single quarry the different beds may vary considerably. Some quarries even supply a combination of entirely different rocks such as limestones and sandstones. Shropshire is a county noted for its varied geology, a fact which becomes increasingly apparent as its building stones are examined.

2
Grinshill Stone

In order to understand the nature of Grinshill stone it is essential to have some insight into the environmental conditions that prevailed over much of north Shropshire during the Triassic Period of some 200 million years ago. Grinshill Hill is now located approximately fifty-two degrees north of the equator and therefore in the path of the many rain-bearing depressions which are swept along in the westerly wind belt. In complete contrast, in Triassic times it was located much further south (probably somewhere between twenty and forty degrees north of the equator) and experiencing the kind of climatic conditions now found in the trade wind belts of the world, notably the North African Sahara. In the Triassic Period Britain was an integral part of the Pangaea supercontinent which was later to fracture and drift apart in huge continental blocks. The one containing the future British Isles was destined for a northerly route. For the time being, however, it experienced aeolian conditions in arid and semi-arid latitudes.

Within the arid environments there were local variations of landscape including extensive featureless plains, dune belts, ephemeral and saline lakes and deep valleys which were cut by sudden, destructive torrents racing down from residual uplands lashed by infrequent desert rainstorms. In such conditions the sandstones, conglomerates and salt deposits of the English Midlands were deposited.

The superior building freestones, now found in massive beds of Lower Keuper age (a subdivision of the Triassic Period), most probably result from the steady accumulation of sands in lakes which developed in a less arid phase of the Triassic. The sand grains are rounded and therefore of aeolian origin, but they are well sorted and occasionally current-bedded so they must have been deposited in relatively calm waters. The thinly bedded Waterstones which accumulated on top of these thick sandstone beds indicate a change of climate, for they consist of alternating shallow layers of shale and sandstone which represent more disturbed conditions when the lakes dried out into temporary mudflats. These can now be recognized by the fossilized animal tracks and mud cracks found on the upper surfaces of the shales. The Keuper Marl above the Waterstones consists of fine silica dust particles derived from the windswept desert plains, in addition to slightly calcareous clays which accumulated in very shallow waters. Here are found the salt deposits for which the Midlands are renowned.

In the closing years of the nineteenth century the sandstone quarries at Grinshill were supplying dressed blocks of superior quality freestone measuring twelve feet by five feet by five feet, extracted from the best building stone layers which reached thicknesses of fifteen feet. Beneath, lay thick beds of a softer sandstone which was not suitable for prestigious buildings but was extremely useful in the immediate locality for the construction of vernacular premises, churches and chapels. A different quality sandstone was found on top of the freestone beds. This was extremely tough but only three feet thick at most. Known as the Hard Burr, it was ideal for the production of doorsteps, pavings and kerbstones as it was fine-grained and durable and capable of resisting constant abrasion. Because it contained specks of black manganese dioxide it could never be used for facework but was often used as backing and foundations to walls constructed of quality freestone.

The overburden on top of these various sandstone strata was not devoid of commercial interest for it consisted of the Waterstones and the Keuper Marl. In the Grinshill district the Waterstones contained thin layers of argillaceous sandstone with concentrations of mica plates which rendered the rock so fissile that it was an ideal source of roofing stones. It was known locally as the Flag Rock because it was also a very useful source of flagstones and hearthstones. The marl could be used for bricks.

During the nineteenth century the roofing 'slates' lost much of their former importance but the other stones supplied by the Grinshill quarries were marketed at ever-increasing distances with the availability of rail transport. New markets were found in the North-west, the Midlands and the London region, where the quality freestone was much in demand for the construction of municipal buildings. The less prestigious stones found a ready market in Shropshire itself, usually for a very wide range of domestic, agricultural and ecclesiastical premises of various architectural styles. The demand for the best Grinshill stone continued within the county for the construction of the more exclusive buildings, but the tendency to use it as dressing stone to other materials increased in the second half of the nineteenth century. Most of Shropshire's outstanding Grinshill stone premises were erected well before the railway system had evolved.

For several centuries blocks of Grinshill stone had been laboriously hauled out of the quarries and conveyed to the building sites by teams of horses drawing carts and waggons across fields and along trackways and roads. Wherever streams and rivers were available for at least parts of the journeys, they were invariably used. It is thought that blocks of Grinshill stone may have been floated down the Roden, Tern and Severn to reach the abbey construction site at Buildwas in the twelfth century. The Severn must have offered some scope for water transport in and around the Shrewsbury and

Moreton Corbet Castle — a sixteenth-century facade of ashlared, moulded and carved Grinshill stone.

Bridgnorth districts and on occasions as far downstream as Gloucester. Horse-drawn vehicles were, however, needed to convey the stone to the watercourses and from these to the building sites. Records show how the keystone for the centre arch of the English Bridge at Shrewsbury was hauled from the Grinshill quarries by a team of twelve horses in 1773 with no indication of water transport being used at all. Broad wheel waggons were essential to prevent the heavy loads from severely damaging the roadways and possibly sinking into poorly constructed trackways. The fact that Grinshill freestone was used in the Shrewsbury area from medieval times shows just how much this stone was regarded as superior to others. For a time its use was restricted by the lure of more easily available stones which were quarried in and close to the town, but in the eighteenth century it became pre-eminent as Shrewsbury's building stone.

The architectural appeal of Grinshill stone is demonstrated by several sixteenth and seventeenth century buildings such as Moreton Corbet Castle, Shrewsbury School (now housing Shrewsbury Library) and the Old Market Hall, Shrewsbury. It was during the eighteenth century, however, that the full potential of Grinshill freestone was most appreciated, for it was a stone absolutely ideal for the restrained classical style of architecture then in vogue. The large blocks of fine-textured, pale-coloured stone were suited to the production of any ashlared facework with a smooth finish. Columns could be turned with relative ease and mouldings accomplished with fine detailing. George Steuart exploited these attributes of Grinshill freestone at St. Chad's, Shrewsbury, All Saints, Wellington, and Attingham Hall at Atcham. John Gwynn made excellent use of the stone in bridge construction at Atcham and on the English Bridge in Shrewsbury; Carline and Tilley were responsible for the Welsh Bridge in the same town. These engineers/architects concerned with the design and construction of such major bridges spanning the Severn were appreciative of the stone's architectural appeal as well as its engineering strength.

The geographical location of the Grinshill quarries in part explains their success in the eighteenth and early nineteenth centuries for they were situated within tolerable reach (even for rudimentary forms of transport) of the county town, where the demand for imposing structures was concentrated. Almost as significant, they were placed in the most productive agricultural region of Shropshire, namely the North Shropshire Plain, where large estates and

Part of the seventeenth-century additions to Shrewsbury School Buildings faced with ashlared and carved Grinshill freestone.

The late sixteenth-century Old Market Hall in The Square, Shrewsbury, with the first floor supported on Tuscan columns. The mullioned and transomed windows and the unusual parapet are echoed on Shrewsbury School Buildings.

prosperous farms were often enhanced by impressive domestic premises. Additional demand for the stone came from the developing coalfield district of eastern Shropshire where wealth-creating industries were encouraging the growth of new settlements, some requiring new ecclesiastical and municipal buildings of style. The Grinshill location was also favoured by the coming of the railways in the mid-nineteenth century as the bulk of the major trunk routes crossed the North Shropshire Plain. The quarries were only one and a half kilometres east of Yorton station on the LNWR Shrewsbury to Crewe line which connected into the national rail network to give access to the industrial cities of north and central England, the North Wales resort towns and London (Fig. 5).

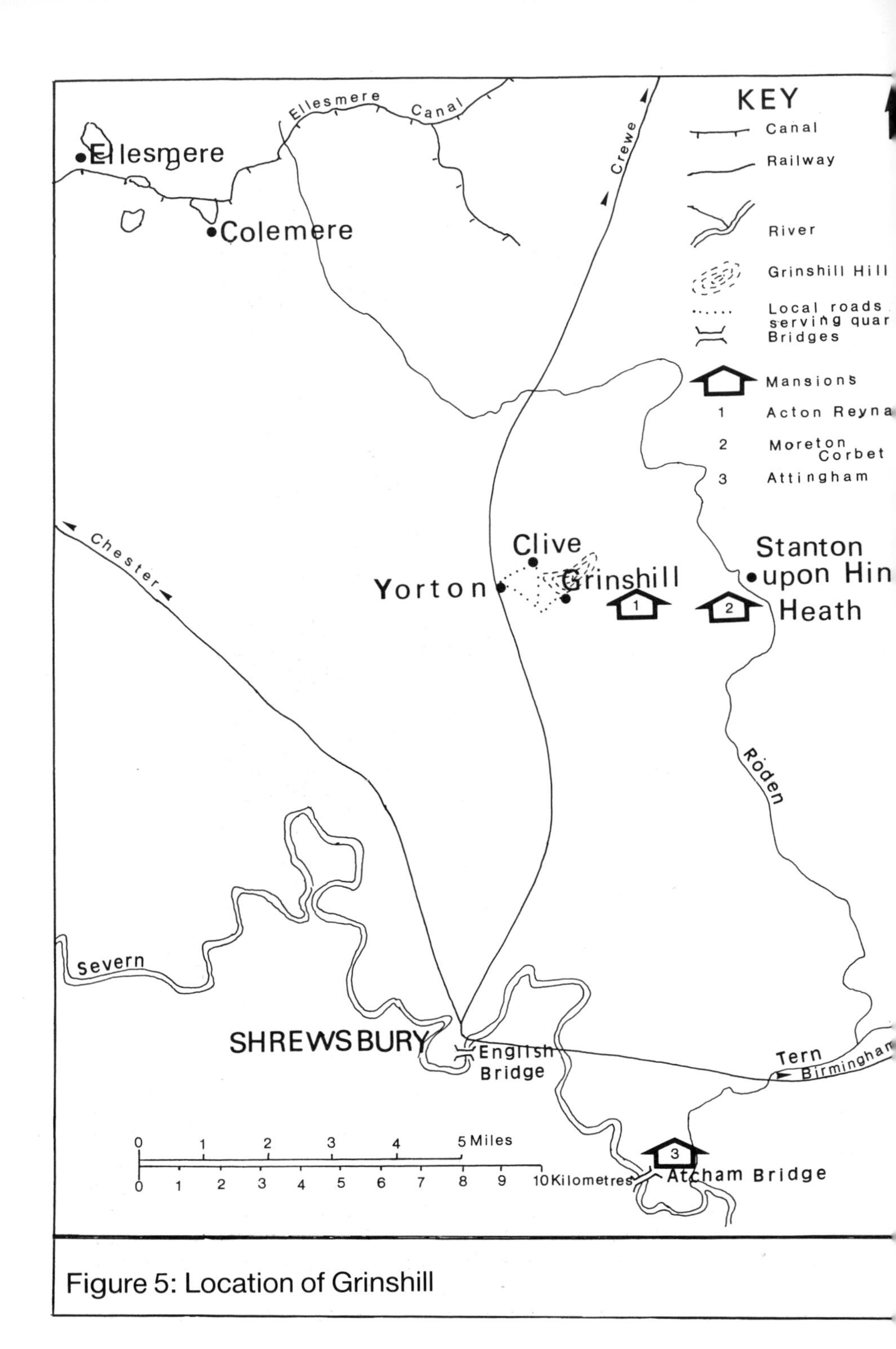

Figure 5: Location of Grinshill

St. Chad's, showing details of the rusticated lower storey; round-headed windows; Ionic pilasters and the parapet punctuated with balusters.

Although the geographical location of any commercial enterprise contributes to its success, the quality of the product is crucial to its appeal. This is especially true of stone which, being a natural as opposed to a manufactured material, is entirely dependent on its geological endowment for its intrinsic characteristics. The critical significance of purely geological considerations is borne out by the remoteness of some difficult and even dangerous quarrying developments. Fortunately at Grinshill, a stone of superior calibre could be extracted and transported without undue difficulty from a hilly site located within a populated lowland region of Shropshire. This favourable combination of circumstances established Grinshill freestone as the premier building stone in the most economically advanced parts of thc county and eventually provided it with the scope to compete successfully with comparable building

Detail of a fluted basement window head at Attingham. Note the square-sunk margins around the stone blocks which produce the 'smooth rustication' effect intended to add weight and dignity to classical, monumental architecture.

Atcham Bridge designed by John Gwynn and completed in 1776 in Grinshill freestone. A classical eighteenth-century bridge with a pronounced rise in the centre. Note the rusticated arch voussoirs and soffits combined with a plain parapet.

stones from other counties of Britain during the building booms of the nineteenth century.

The geological background to the Grinshill quarrying industry has been lightly sketched in with reference to the amassing of sandstones during the desert phases of the distant past. It now becomes necessary to focus attention on the particular characteristics of the stones extracted at the quarries in relation to the actual structures into which they were fixed.

It is essential that a stone required for impressive architecture should be available in large blocks of uniform texture and colour. Eighteenth-century churches and mansions of classical design were especially dependent on ashlared freestone to provide for the refined masonry work demanded by the restrained and dignified architecture. George Steuart used large facing blocks of Grinshill stone in the construction of both St. Chad's and Attingham Hall in the late eighteenth century. The 'Gothick' nave of St. Alkmund, Shrewsbury, built by Carline and Tilley, was also executed in large blocks of Grinshill freestone in contrast with the small blocks of stone used on the medieval tower.

The fashion for monumental walling masonry continued into the early decades of the nineteenth century. St. George, Frankwell, which was designed by the Shrewsbury architect Edward Haycock as a typically plain lancet style Commissioners' church of the 1830s, contains some large blocks of Grinshill freestone which now appear to overwhelm the scale of the building on its

John Gwynn was involved in the design of the English Bridge in Shrewsbury. This classical eighteenth-century bridge was originally hump-backed and had to be lowered and widened in the early 1920s.

The Welsh Bridge, Shrewsbury, was built of Grinshill freestone by Carline and Tilley in the 1790s to replace the medieval bridge with its gate-tower.

cramped site. Throughout the nineteenth century, as towns expanded and required an increasingly varied range of administrative and commercial premises, large dimension blocks of quality stone were needed for the facades, if not the entire walling masonry of imposing buildings. Often the stone was used as facing to less prestigious materials since good freestone was relatively expensive. Nevertheless, the demand for generously proportioned facing blocks to suit the grandiose municipal architecture maintained employment at the freestone quarries until the early decades of the twentieth century when rival materials challenged the supremacy of natural stone.

Weighty and durable stones were also essential in most engineering structures. Early bridges depended on load-bearing masonry but gradually lighter and more precisely engineered structures incorporated facework of freestones and backing of less visually appealing stones. All had to be sound and of large dimension since the number of joints, the potentially weak elements of a bridge structure, had to be kept to a minimum. This was especially the case for any piers and abutments which had to withstand the full force of river currents.

In the nineteenth century great demand for quality building stone was placed on the freestone quarries by railway architects engaged in the construction of impressive station halls for the fiercely competitive railway companies. Grinshill stone was chosen for stations at Shrewsbury, Chester, Crewe, Gloucester and even at Cheltenham, which was close to quarries supplying limestones similar to the famous Bath stone. The railway engineers responsible for the construction of massive viaducts, bridges and embankments did not necessarily require the top quality architectural stones used on the station facades, but the stones had to be sound and durable and were often obtained from the same quarries. Waterworks and harbour engineers also had need for sound, weighty and resistant stone blocks since these were relatively easy to set down in water. Quarries capable of supplying large dimension blocks were at a premium in the nineteenth century.

The Grinshill quarries, despite several centuries of exploitation, were still producing large dimension blocks throughout the nineteenth century. However, when the English Bridge was being reconstructed in 1926 the famous Portland limestone quarries of Dorset supplied the stones for the three keystones of the centre arch because the Grinshill quarries were no longer able to provide the required sizes, although for many centuries previously, massive blocks had been extracted from the best building stone beds which, as already noted, were up to fifteen feet thick. This bounty of fine-grained, even-textured and pale-coloured sandstone was the legacy of the steady accumulation of desert sands during the Lower Keuper phase of the Triassic.

The relatively quiet sedimentation, however, was just one episode in the formation of the Grinshill sandstone. No part of the earth's crust is forever entirely immune from disturbances which relate to the fact that the solid crust is merely 'floating' on molten material in which currents rise and fall according to temperature and density differences at depth. The rocks at the surface respond to these subsurface movements by fracturing and shifting laterally so that molten rock, fluids and gases move in. The invasive materials either seep through the rocks or blast their way through to reach the surface in volcanic eruptions. In either case the existing rocks are substantially altered by heat, tension or compression, and in the second instance by shattering.

Fortunately, nothing too drastic happened to the deep accumulations of sandstones in north Shropshire. Post-Triassic earth movements actually benefited the quarrying activities of the far distant future by causing blocks of the crust to fracture along fault lines and tilt at gentle angles in such a way that the relatively tough sandstones emerged as low hill masses after prolonged erosion (Fig. 6). The possibility of quarrying into a hillside with the added bonus of transporting the stone blocks downgrade is a great advantage in stone extraction. The sandstone hills of north Shropshire, which stretch from Nesscliffe in the west across to Marchamley and Hodnet in the east, show much evidence of quarrying.

The hilly relief, together with the near horizontal bedding, made it possible to quarry the building stones vertically at Grinshill and then transport them downgrade in carts to the stoneyards for cutting and dressing. Further movement by horse-drawn waggons to the local roads and later the LNWR Shrewsbury to Crewe railway line was also downgrade.

Before the main building stone beds were reached, the overburden had to be removed and the Flag Rock and the Hard Burr quarried. Only then could the freestone quarrying commence. First the top surface of the upper building stone bed was carefully marked out into rectangles which were then

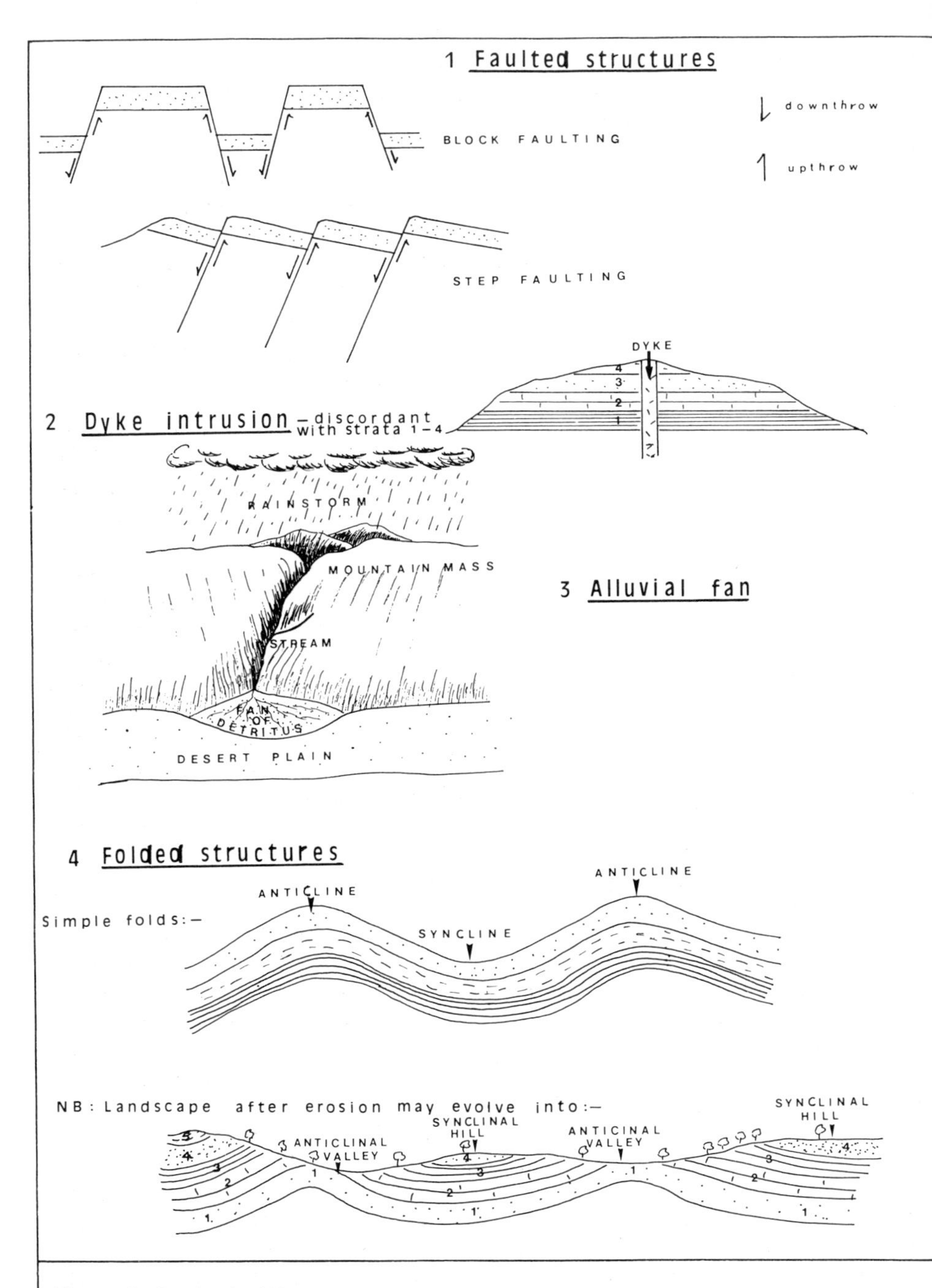
1 Faulted structures
downthrow
upthrow
BLOCK FAULTING
STEP FAULTING
DYKE
4
3
2
1
2 Dyke intrusion – discordant with strata 1–4
RAINSTORM
MOUNTAIN MASS
STREAM
FAN OF DETRITUS
DESERT PLAIN
3 Alluvial fan
4 Folded structures
Simple folds:–
ANTICLINE
SYNCLINE
ANTICLINE
NB: Landscape after erosion may evolve into:–
ANTICLINAL VALLEY
SYNCLINAL HILL
ANTICINAL VALLEY
SYNCLINAL HILL
3
4
3
2
1
1
4
3
2
1
1
3
4
2
1

Figure 6: Geological Features

laboriously cut around with the aid of hand-operated jadd picks. The grooves or 'gutters' thus created were approximately nine to twelve inches wide and three to four feet deep. Wedges inserted into the grooves were manipulated until each individually marked out block could be loosened from its lower bedding plane and detached from the bedrock. It was then lifted out by means of pinch bars which gripped the side of the block ever more tightly as it was hoisted. Such traditional methods of quarrying persisted in the freestone quarries of north Shropshire and neighbouring Staffordshire throughout the nineteenth century, despite the invention of channelling machines. Manual methods were regarded as less damaging to quality freestones.

If post-Triassic faulting movements indirectly contributed to the relative ease of exploiting hilly outcrops of stone, localized igneous activity in the vicinity of Grinshill during the Tertiary period had more direct beneficial effects on the calibre of the sandstone. The intrusion of a pophyritic dyke (Fig. 6) and the occurrence of barytes-rich mineralization altered the original desert sandstone into the paler and harder version destined to become the esteemed 'white' Grinshill freestone keenly sought by architects, masons and builders during several centuries of exploitation.

However several buildings in the Shrewsbury district contain red sandstones from the many quarries scattered across the Triassic sandstone hills of north Shropshire. The red sandstone was far commoner than the 'white' variety. There were quarries at Little Ness, Hopton, Ruyton-XI-Towns, Myddle, Harmer Hill, Clive, Grinshill and Weston. St. Mary's and Holy Cross Abbey in Shrewsbury contain some of this Triassic red sandstone in addition to the locally quarried Carboniferous sandstones referred to in Chapter 6. Thomas Telford's Montford Bridge, which spans the Severn to the north-west of Shrewsbury, was constructed of red sandstone quarried on Nesscliffe Hill which is only six kilometres to the north-west, along the Shrewsbury to Oswestry section of the Holyhead road. The building contract of 1790 insisted on the use of broad-wheeled vehicles for cartage down from the quarries. The blocks must have been massive.

Telford's choice of the red sandstone was probably inspired by the simple expediency of using the local Nesscliffe outcrop. By the late eighteenth century the pale-coloured sandstones obtained from Grinshill were usually the preferred choice of architects and engineers. For example, two contemporary constructions, the Welsh Bridge in Shrewsbury, built by Carline and Tilley, and St. Chad's church by George Steuart, relied on best Grinshill freestone for their effective classical designs.

This preference for white Grinshill stone continued into the early decades of the nineteenth century. Architects responsible for the design of several prestigious buildings erected in the centre of Shrewsbury perhaps understandably followed the precedent set by two long-established buildings of repute, the Old Market Hall and Shrewsbury School. Both displayed the advantages of using Grinshill freestone for finely executed carving and smooth ashlar masonry. Despite the inevitable wear and tear apparent in the early nineteenth century, these familiar buildings must have offered proof that the pale versions of Grinshill stone weathered much better than the red sandstones quarried from the Triassic sandstone hills to the north of the town.

Edward Haycock designed two severely classical buildings for conspicuous sites in Shrewsbury. The Royal Salop Infirmary, constructed 1826-30 in St. Mary's Place, contrasted boldly with the nearby medieval church which had patchworth masonry fabricated from various red and yellow sandstones. The Infirmary was built of uniformly pale Grinshill freestone which suited the plain facade with its Doric portico and giant pilasters set against the smoothly finished ashlared walling masonry. The Music Hall, constructed 1839-40, was the second of Haycock's austere designs, with a giant fluted Ionic portico with pediment fronting the plain five-bay facade.

Robert Smirke's Shirehall of 1834 was in a more lively Italianate style with an emphatic cornice resting on carved modillions and with more elaborate

The severe Greek Doric portico of the Salop Infirmary, Shrewsbury, projects from a plain ashlared Grinshill stone facade.

St. Michael, Chirbury — the south aisle wall of random-coursed Whittery rock combined with Shelvock freestone dressings.

All Saints, Baschurch — detail of walling masonry showing how weathering has penetrated deeper than the mason's herringbone tooling.

Cross-stratified Dune Sandstone outcropping alongside the Bridgnorth to Kidderminster road at Quatford.

Weathering accentuates the cross-stratified bedding of the friable local red sandstone used for the walling of St. Leonard, Bridgnorth.

Tufa on St. Mary Magdalene, Quatford.

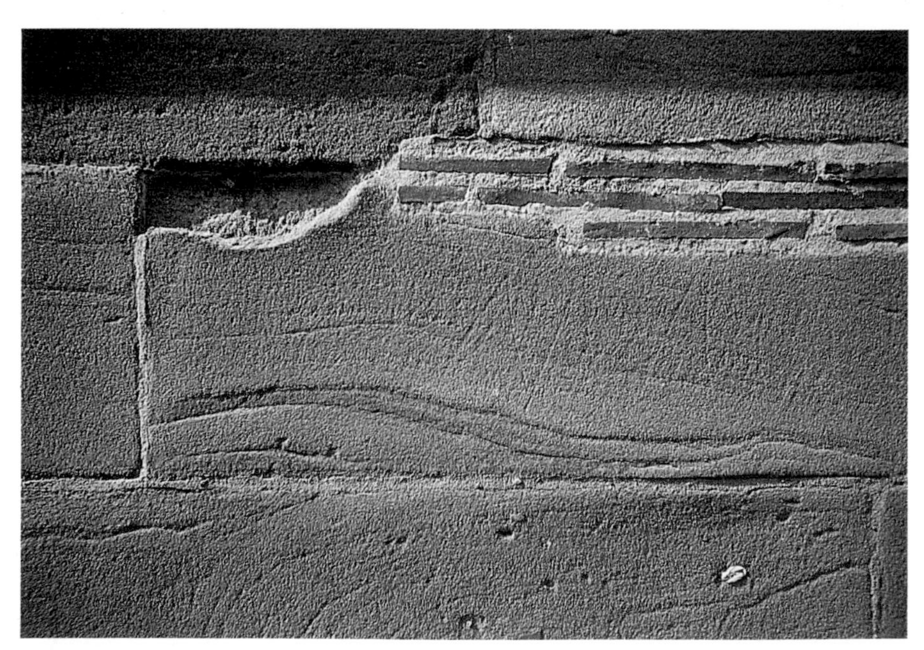

Pebble Bed sandstones used in the late nineteenth-century masonry on St. Mary, Market Drayton. Note how tiles have been inserted to plug the weathered-out clayey lenses in the sandstone.

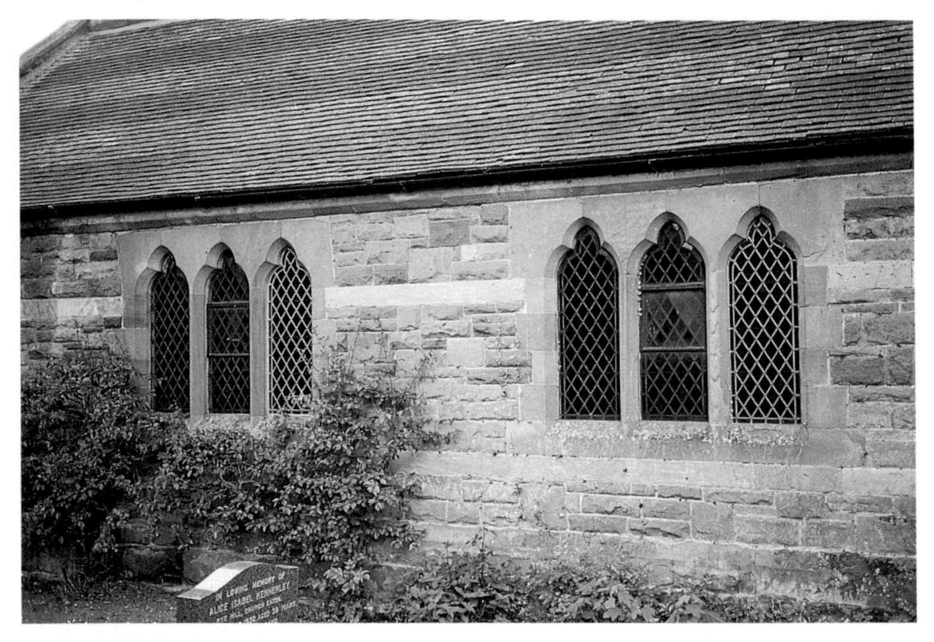

Rock-faced, random-coursed Pebble Bed sandstones combined with freestone bands on the walls of St. Michael, Chetwynd.

Farlow Sandstone on the walls of St. Giles, Farlow. Note the thinly bedded, easily weathered block in the centre.

Shrewsbury Station, altered in the early twentieth century, following Penson's original Tudor-Gothic design which relied on ashlared, moulded and carved Grinshill freestone for its architectural effects.

detailing on the windows, some of which carried moulded pediments. Each of these three buildings displayed some measure of classical restraint and therefore contrasted with the Old Market Hall which had a vivacity of outline imparted by the ornamental parapet and the arcaded lower storey. Between them, these buildings reflected the capacity of Grinshill freestone to accommodate both smoothly dressed ashlar walling and lively moulding and sculpturing.

It is no surprise that T K Penson chose the same stone for the facade of Shrewsbury Station which was constructed in the mid-nineteenth century. A six-inch outer skin of ashlared Grinshill walling stone formed the plain backdrop to such decorative elements as the richly carved base of an oriel window, the ornamented string courses and the sculptured heads which served as stops to the window labels. The Tudor-Gothic design with its tall central tower was intended to impress.

The close-grained, fine-textured Grinshill stone was relatively easy to work. When freshly quarried it was 'soft', but it hardened sufficiently on exposure to the atmosphere to preserve any fine masonry details. It was therefore an ideal stone for sculpture, both for internal and external work.

A memorial to John Corbet who died in 1817 was worked by Carline and Son and installed in Battlefield church, which, as the commemorative church for the Battle of Shrewsbury, dated back to the fifteenth century. The monument was an extremely elaborate Gothic design with mouldings, crockets and finials, exemplifying Grinshill stone's capacity to support

The Greek Doric column of Grinshill freestone supporting the statue of Lord Hill, London Road, Shrewsbury.

intricate carving. This attribute must have been exploited from an early date, for when George Edmund Street was engaged in restoring the medieval church of St. Michael, High Ercall, in the 1860s, he insisted that the replacement caps of the Transitional arcade should be carved by a skilled mason from ample-sized blocks of Grinshill stone in accordance with the precedent set by the medieval masons.

The comparative ease with which Grinshill stone could be carved made it an obvious choice for masons involved in working window tracery. For this reason it was frequently used in church premises otherwise built of different stone and even, on occasions, of brickwork. When St. Michael, Smethcott, was rebuilt 1849-50 of rough rubblestone from a nearby quarry on the Precambrian gritstones of the Longmynd, Grinshill stone was introduced for the window tracery although the window surrounds were cut out of the tough Ordovician Hoar Edge Grit. At St. George, Rodington, also rebuilt in the mid-nineteenth century, the architect, Ewan Christian, combined Grinshill stone window tracery with red brickwork.

Despite the initial ease of carving Grinshill freestone, any tracery for which it was used had to have good wearing potential to merit the delicate workmanship entailed. Fortunately, the faith of the masons was usually justified. When R H Carpenter was engaged on the restoration of St. Peter, Wrockwardine, in 1887, he advised that the east window of Middle Pointed date should be retained since only minor repairs to the cusps and tracery were required. The original window had been carved from the best quality Grinshill stone.

A more arresting testimonial to the weathering potential of Grinshill stone is the fluted Greek Doric column supporting Lord Hill's statue which is located in a prominent and unprotected position at the top of the London Road in Shrewsbury. Ninety-one feet high, it stands on a pedestal featuring four lions couchant carved out of massive blocks of Grinshill stone. The figure of Lord Hill is not, however, of the freestone but of a purely artificial stone manufactured by Coade and Sealy, a firm well versed in the production of fabricated stone by the early nineteenth century when the monument was erected to commemorate Lord Hill's military exploits. Edward Haycock was responsible for the initial design but it was completed by Thomas Harrison of Chester. Both architects were exponents of the austere classical style.

Mention has already been made of the high silica content of Grinshill stone which improved the capacity of quarried blocks to withstand the rigours of diverse environments. They were much in demand by those nineteenth-century architects engaged in the construction of municipal buildings in Liverpool, Birmingham and a number of smaller smoke-begrimed industrial towns. Occasionally, the entire facade of a civic building was executed in the freestone. This was the case at Newport, Monmouthshire, where the Free Classic style entrance front of the New Town Hall on Commercial Street was worked in Grinshill stone. The joint architect, Thomas M Lockwood, had become familiar with the stone whilst involved in the renovation of Smirke's Shrewsbury Shirehall in the early 1880s. He used it again on the Grosvenor Museum in Chester which was built in the same decade. This time, however, the stone was combined with Ruabon brick and was used to highlight the cornices and pilasters. Such limiting of stone to the dressings became an established practice in the late nineteenth century as freestone became relatively expensive and tantamount to a luxury building material. Polychromatic effects were also popular as a design feature with pale-coloured stones providing textural and colour contrasts to shiny red bricks. The combination of Ruabon brick and Grinshill stone can be found in Shrewsbury on the nave of Holy Trinity, Belle Vue, which was rebuilt in 1886-7.

The high proportion of silica in its chemical content also gave Grinshill stone a fair measure of protection against the ravages of salt-water spray. Hence its use in a variety of premises erected along the coastlines of north and central Wales in the second half of the nineteenth century. It was combined with the local blue slate stone in the construction of the Queen's Hotel at Aberystwyth in the 1860s. As well as alleviating the oppressively dark colour of the slate, the Grinshill freestone was needed for the precisely cut quoins and other dressings. These could not have been produced from the fissile slatey local rock. The Ladies' Hostel and the University Buildings at Aberystwyth also featured Grinshill freestone, in these instances as the main walling material.

At the end of the nineteenth century the waterworks engineers employed on the construction of the Elan Valley reservoirs exploited the capacity of

George Steuart's smoke-begrimed classical church of All Saints, Wellington, showing the west front of Grinshill freestone with giant Tuscan pilasters, decorated pediment and semi-circular upper windows set in blank arches.

The east end of All Saints, Wellington, with 'Georgian-style' shallow apse.

Grinshill stone (by reason of its high silica content) to withstand the effects of acidic mountain waters. This was a very specialized use of a stone that was primarily of architectural merit, but not without utility as an engineering stone, since it had sufficient crushing strength and specific gravity for most engineering uses, including bridge construction.

This brief review of the characteristics of Grinshill stone explains its different roles throughout several centuries of exploitation by architects, engineers, masons and sculptors. In the early days, its use was concentrated in the Shrewsbury district with some extension southwards along the Severn valley. By the eighteenth century its market area had encroached into other parts of Shropshire, notably the developing coalfield area of eastern Shropshire. Here, for example, 'white' Grinshill freestone had been used by George Steuart on his classical style church of All Saints, Wellington, which was completed in 1790. All Saints lost much of its pristine appearance after the construction of the nearby railway in 1838. Nevertheless, the stone showed a remarkable degree of resistance to the smoke pollution, a point not lost on the nineteenth-century architects involved in the urban developments of the North-west and of the Midlands. By the end of the nineteenth century the market range of Grinshill stone was nationwide.

It was during the eighteenth century and the early decades of the nineteenth century that the most lavish use of Grinshill freestone occurred on many prestigious, stone-built classical buildings and bridges. As the architectural style demanded neatly cut ashlar blocks and precisely turned columns and balusters, only the best quality freestone would suffice.

Blemishes were all too conspicuous against the clean and austere lines of the architecture.

From the mid-nineteenth century on, the combined effects of economic, technological and aesthetic influences intensified the use of Grinshill stone in its role as a dressing stone to a variety of other materials. The combination of brick and freestone became commonplace on a diversity of buildings ranging from the most sophisticated of municipal edifices to the larger domestic premises housing middle-class families. City architects were the main exponents of structural polychromy which depended on the admixture of materials of different textures and colours. Freestone invariably emphasised the focal elements of the architecture and featured prominently on the main facades. For the Bishton and Fletcher premises in Albion Street, Newhall, Birmingham, T Tadman Foulkes produced an Italianate design with red bricks embellished with Grinshill stone dressings on the sills, strings and main cornice. This reflected the achitectural mode of many industrial and commercial buildings erected throughout the nation.

The use of Grinshill freestone declined in the twentieth century as artificial stone and aggregates became fiercely competitive with the natural stones quarried throughout Britain. The Grinshill quarries had to be reopened for the restoration of the Shrewsbury School Buildings in the early 1980s. This event marked approximately eight centuries of exploitation of the stone. During this long span of time the freestone had been used with distinction by a few architects and engineers of renown and more generally, though not necessarily less successfully, by local builders and masons solely concerned with the provision of buildings for the rural communities of north Shropshire.

3
The New Red Sandstone of North Shropshire

Grinshill stone is just one of the many sandstones of North Shropshire that are found amongst those Triassic and Permo-Triassic rocks (alternatively named New Red Sandstone rocks) which cover large tracts of the northern and eastern parts of the county.

The best building sandstones are usually found amongst the Triassic Lower Keuper Beds which outcrop in a series of hills stretching from West Felton in the west to Hodnet in the east. These hills are the sites of many quarries which in the past supplied red sandstone blocks for the construction of local churches, chapels, farmhouses, cottages and the occasional country mansion. Some of the red sandstone was incorporated into the fabric of churches in Shrewsbury, for example the Abbey church and St. Mary's, and also in major bridges such as Telford's Montford Bridge, but generally its use was very localized and connected with the construction of vernacular buildings for the rural communities of the Shropshire Plain. No other stone of this region equalled the reputation of Grinshill stone and measured up to the competition of quality stones from other regions of Britain, but a few select quarries supplied stones which, being superior to the commonplace red sandstones of the district, were sought out by both local and national architects engaged on more important commissions within the county.

One such quarry was that at Shelvock, located on the Triassic Lower Keuper Sandstone mid-way between West Felton and Ruyton-XI-Towns. It was a deep quarry providing thick beds of good quality red and paler-coloured sandstones. Because of its pleasing texture and colour, the quarried stone was valued for ornamental work and for the architectural highlights on buildings composed of other materials. For example, it featured in the stone Ionic columns embellishing the facade of Woodhouse, a brick house designed by Robert Mylne and built in 1773-4 some five kilometres north of the quarry.

The striking colour of the stone appealed most particularly to Victorian architects anxious to achieve the polychromatic effects in architectural design which were publicized by writers such as John Ruskin in the mid-nineteenth century. Ruskin's *Stones of Venice* had great influence on the architectural fraternity, encouraging a fashion for combining materials of various textures and colours on a single building, sometimes with bizarre results. In other instances, however, the colour contrasts were muted. Edward Haycock, a Shrewsbury architect, combined red Shelvock stone with buff Cefn stone quarried near Ruabon, on the fabric of both Oswestry School Chapel, completed in 1863, and Welsh Frankton church, built 1857-8.

It was usually on town churches that the fashion for colour combinations was flaunted. On All Saints, Castlefields, Shrewsbury, Shelvock stone appears with Grinshill, Red Hill and Bristol Blue Pennant stones, the dominant stone being Red Hill which was quarried out of Carboniferous sandstones outcropping to the south-west of Shrewsbury. This very deep-

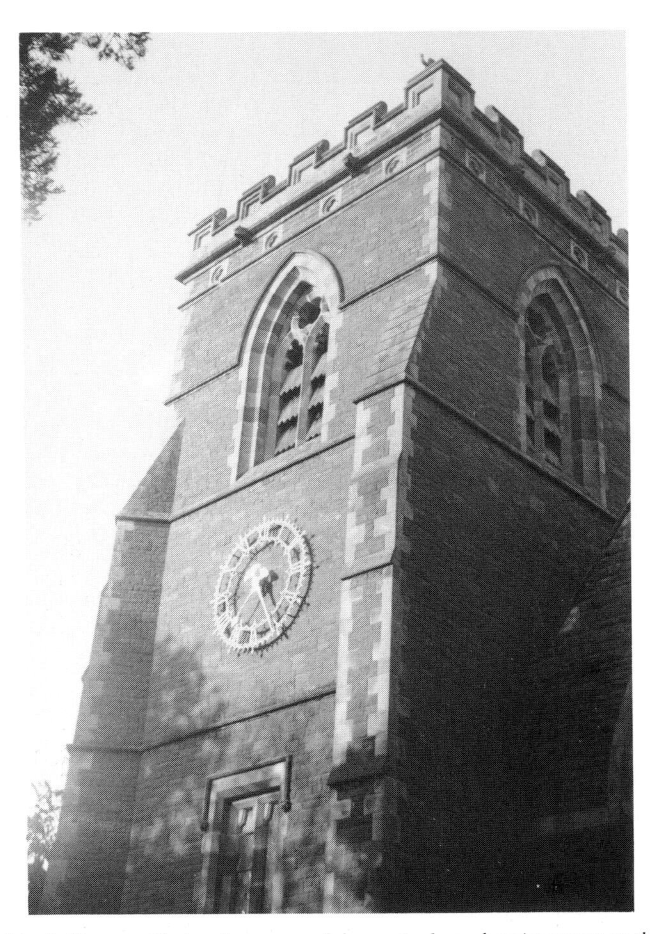

Holy Trinity, Meole Brace — Shelvock stone used decoratively as dressing stone on the tower.

toned Red Hill stone was also used by Edward Haycock for the main walling masonry of Meole Brace church, built 1867-8. He chose the paler Shelvock stone to accentuate the door and window openings, the buttress quoins and the tower parapet.

The fine texture of Shelvock freestone was a quality much appreciated by Shropshire architects working with some of the tougher stones quarried on the western borders of the county. For example, Haycock combined Shelvock stone with the inherently tough Llanymynech limestone on the fabric of St. John the Evangelist, Maesbrook, which was constructed in the 1870s. More telling is the use he made of the comparatively easily worked Shelvock freestone for the restoration of St. Michael, Chirbury, in 1871. During this major restoration, which cost some £1,700, the south aisle wall was completely rebuilt in a very tough Ordovician rock quarried locally in Whittery Quarry. As this could only be roughly squared, it was laid to random courses and set in thick mortar comprising blacksmiths' ashes, sand and lime. The faces of the random-sized stones were left rough in contrast to those of the Shelvock stones which were smoothly rubbed down. The Shelvock freestone was essential for the working of the window surrounds and tracery, all of which was very close-jointed and neatly dressed. The ease of tooling the Shelvock stone in this instance was more important than the colour contrast it introduced.

A E Lloyd Oswell, an Oswestry architect, also made good use of Shelvock in addition to Grinshill stone for the dressings on the church of Holy Trinity, Bicton, which was built in the 1880s. The main walling stone was the extremely tough, coarse-textured and lumpy Alberbury Breccia, a Permian rock quarried some thirteen kilometres west of Shrewsbury in the hilly tract between Alberbury and Cardeston. Although a very serviceable rock, the breccia was impossible to carve and mould so that freestones were needed for the finer masonry work.

Of the numerous sandstone quarries in North Shropshire a few merit attention because of their special connection with more prestigious properties. Some of these quarries were opened up specifically for an individual commission. Ruyton Manor, built in the 1870s just west of the settlement of Ruyton-XI-Towns, was designed in the typically grand and somewhat ponderous castellated style much favoured at the time. The large consignments of stone which were required in the construction of the mansion warranted the laying of a light railway which ran from a nearby red sandstone quarry on Grug Hill down to the building site. According to a contemporary report in the *Builder*, it was the 'ordinary workmen of the district' who built the sizeable edifice complete with its machicolated and crenellated high central tower — all constructed without the security of external scaffolding. Such men were more accustomed to the task of erecting small cottages and chapels in the neighbourhood of Ruyton-XI-Towns and Baschurch, so the dinner they were treated to part-way through the proceedings was a timely incentive. The quarry on Grug Hill was a near neighbour of Shelvock Quarry and another example of the exploitation of the Lower Keuper sandstones of the Triassic System.

There were many others. The fact that several of these quarries became the source of material for properties designed by leading nineteenth-century architects suggests the good reputation that these sandstones had. Sir Aston Webb designed Yeaton Peverey for an illustrious local client, Sir Offley Wakeman, in the 1890s. The materials were choice, the architecture a conventional interpretation of the currently fashionanable Jacobean style. Best English oak was used for the half-timbering, green Westmorland slates for the roof and bricks and stone for the walling. The stone was essentially local red sandstone. It came from a quarry leased from Sir Rowland Hunt of Boreatton Hall and located higher up the valley of the River Perry to the north of Ruyton-XI-Towns. The sandstone was the principal walling material of this extensive house.

Adcote House, situated just over two kilometres distant from Yeaton Peverey in the direction of Little Ness, had been constructed some twelve years previously to a design of Richard Norman Shaw. Shaw was a famous nineteenth-century architect based in London, with both public and private buildings to his credit. During the second half of the century his reputation for country house design grew to the extent that he became a much sought-after architect for clients with deep pockets and good taste. Wealthy industrialists engaged him for the construction of fine domestic properties which reflected their financial status. When Rebecca Darby, the widow of Alfred Darby of Coalbrookdale and herself a member of the wealthy Christy family of London hatters, purchased the Adcote estate in 1868 she required a new house. Shaw was commissioned to design it for her in the 1870s.

Though an exponent of brickwork, half-timbering and tile-hanging on many of his houses, Shaw in this instance virtually confined himself to the use of stone for the external fabric of the new house. The local stone was equal to the exacting requirements of the architect. It came from Little Ness Quarry on the Adcote estate and admirably suited the accomplished design which featured mullioned and transomed windows, clean-cut gables and buttresses. The visible masonry was the expertly contrived facing to a brick lining and concrete infill, for this was a modern structure. The local stone was also used internally and with equal panache on the staircase and in the spaciously impressive hall with its austere but dramatic fireplace.

The stone obtained from Little Ness Quarry resembled the best variety of Grinshill stone, being a pale version of the Lower Keuper building sandstone of North Shropshire. Other quarries nearby on the estate yielded the more familiar red stone. A distinction was made in the use of these estate stones, with the rarer and more finely-textured pale sandstone being reserved for the manor house and the main entrance lodge. The much coarser and more friable red sandstones were liberally used on the estate cottages and farms, especially on the numerous more crudely constructed outbuildings. Bricks made in the estate brickyard were used on larger and more impressive farmhouses.

On the Hawkstone estate, located in the easternmost sandstone hills of the North Shropshire Plain, the deep red sandstone dominates a naturally dramatic landscape that was exploited to the full by Sir Rowland Hill and his successor during the eighteenth century. It was then fashionable for landowners to strive for the most picturesque effects with the prominent siting of parkland features at focal points on their estates. At Hawkstone, a grotto, a red sandstone summerhouse and an obelisk were grouped around ruins of a thirteenth-century castle constructed of the local red sandstone. When the Chester architect, Thomas Harrison, was commissioned for the design of the dowerhouse for the Hill family in the late eighteenth century he devised a castle style suited to the ambience of the park. The Citadel, as the dowerhouse was called, has a more circumspect castellated form than that of Ruyton Towers, which was built in the second half of the nineteenth century with typical Victorian exuberance, but it was constructed of a similar red sandstone, a stone which because of its fiery, aggressive colour was well suited to the castle image.

Although the big houses of the North Shropshire Plain are interesting, each in its own right, it is the vernacular premises that most truly represent the building stone endowment of the region. Most rural buildings are constructed of local materials, so that stone-built cottages and farmhouses closely relate to the rocks over which they are distributed. Since the red sandstones are by far the most common of the Triassic and Permo-Triassic sandstones, they constitute the fabric of most small domestic properties as well as a significant number of village churches and chapels. Large freestone blocks were used on the smallest of cottages and on tiny, simple chapels even though such blocks were disproportionate to the scale of these buildings. Sometimes diagonal tooling disguised the size of a block and gave the masonry a more rustic appearance appropriate to a rural environment.

Sandstone chapel, Bomere Heath.

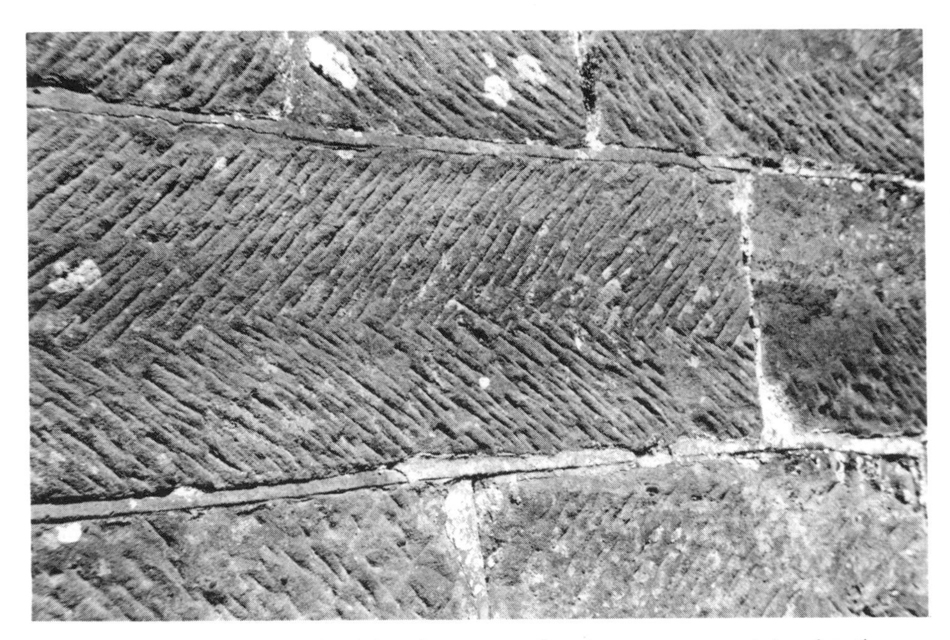

Diagonal tooling — a masonry finish found on many red sandstone cottages and chapels in the north and east of Shropshire.

Many rural premises were, however, of rubblestone construction which is possibly more compatible with their setting (Fig. 7). Much of the rubblestone was essentially quarry waste, the residue left after the removal of the freestone blocks. It was not necessarily inferior material in terms of its inherent durability, but merely small, misshapen fragments of stone. Being much cheaper than the dimension blocks it was widely used by local builders primarily concerned with putting up vernacular buildings for agricultural communities. In many instances the rubblestone masonry has weathered much better than the ashlared freestone masonry, used on those modest premises least likely to be constructed of the best and most expensive freestones.

Generally the red sandstones weather more rapidly than their paler counterparts for, as previously explained, they contain a higher proportion of unstable iron oxides and less of the chemically inert silica which resists the ravages of time. There are many examples of badly weathered red sandstone buildings of medieval date in and around the Shrewsbury district which were mostly built of Triassic sandstones quarried at Nesscliffe, Ruyton-XI-Towns, Baschurch and Harmer Hill. The same quarries were often re-worked for repairs undertaken at subsequent dates. For example, the architect J L Pearson specified the use of Shelvock and Harmer Hill stone in an effort to maintain continuity with existing fabric for the major restoration and rebuilding work undertaken in the 1880s on Holy Cross Abbey, Shrewsbury. The importance of matching ancient stonework was not, however, always

1 RUBBLESTONE

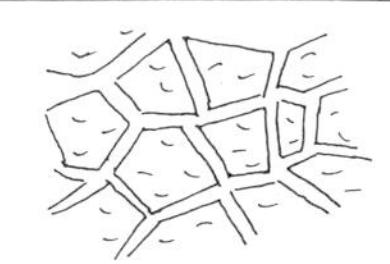

1a: **random**
(uncoursed)

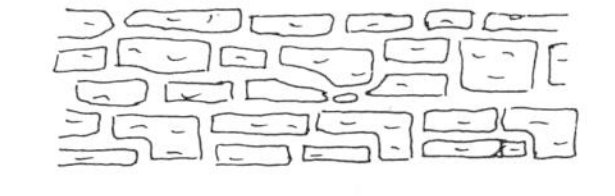

1b: **range rubble**
(random-coursed)

2 SQUARED STONE

(quarry-faced; rock-faced; hammer-dressed)

J - Jumper stone - this large stone may span two or more courses.

2a: **coursed, range work**

2b: **'snecked', broken range work**

3 ASHLAR

(precision cut, dressed, rubbed stonework)

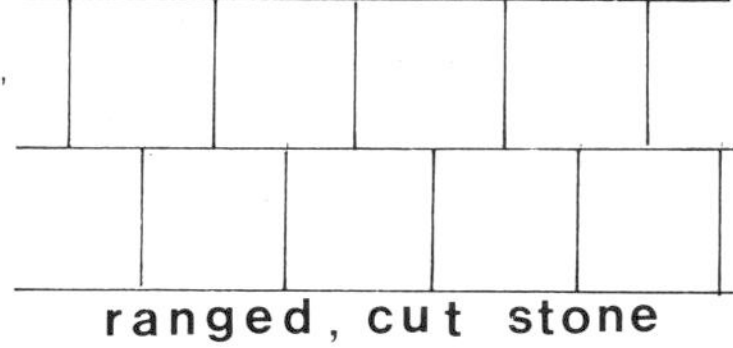

ranged, cut stone

NB Ashlar masonry can be broken ranged or random-coursed but these variants are less suitable for refined, classical architecture.

Special effects:-

1 Rustication

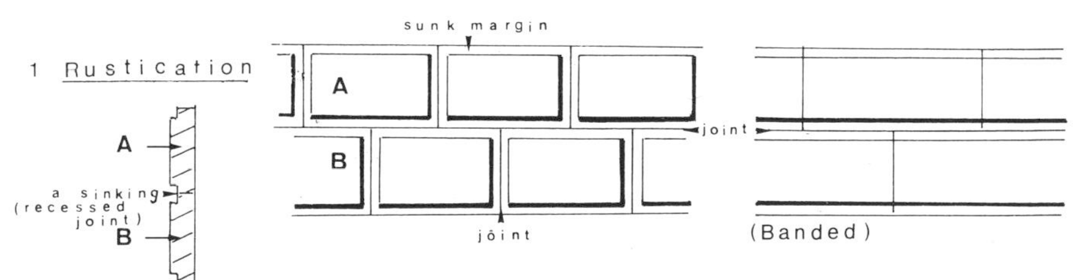

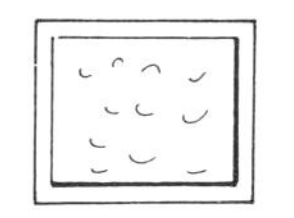

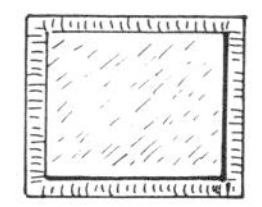

2 Drafted Margins

3 Herringbone Tooling

NB There are gradations between masonry techniques **1 → 2 → 3** and regional differences in terminology.

Figure 7: Masonry

recognized until a comparatively late date in the nineteenth century and some churches in Shrewsbury, such as those of St. Mary and St. Giles, display a patchwork of masonry resulting from the insertion, down the centuries, of the more dependable but paler-coloured sandstones quarried in the Grinshill district.

Several ancient churches built in closer proximity to the red sandstone quarries of the hilly tract north of Shrewsbury also reveal evidence of decay. Such is the case at both Baschurch and Ruyton-XI-Towns. Even more recently constructed churches show considerable evidence of decayed stonework. At St. Luke, Weston, the ashlared blocks on the eighteenth-century tower have peeled away in patches to depths sometimes in excess of two centimetres. This deep spalling of the surface layers is typical of a potentially friable rock which has been smoothly cut and dressed. The faces of the stones thus treated often harden and form a tough, mineral-rich crust beneath which the softer stone may rot away. Even churches built as recently as the late nineteenth century can show signs of such advanced weathering, especially in those parts of the structures most exposed to the elements. Chemical decay and physical disintegration resulting from temperature changes and frost action are particularly active on rocks that absorb moisture readily and release it more slowly, and stones placed where damp persists are extremely vulnerable. Such adverse conditions are found in the stone fabric of many red sandstone churches of North Shropshire.

It would be wrong to assume that all the red sandstones of the Triassic and

Deep weathering of the soft, bright red sandstone of the Baschurch district. This peeling away of surface layers is known as 'spalling'.

Permo-Triassic outcrops of North Shropshire were unreliable as building material. Most quarries could be depended upon to provide better stone, at least from a few select beds. The essential prerequisite for satisfactory masonry work was the careful selection of the best stone for the specific purpose. For his bridge building activities in the northern plain in the late eighteenth and early nineteenth centuries, Thomas Telford, in collaboration with his chief assistant, Thomas Stanton, invariably managed to find the best red sandstone available at the local quarries, but he had the expertise of his early stonemason's training to fall back on. The bridges at Montford and Lee Brockhurst testify to his practised eye. The railway engineers of later date were also sufficiently aware of the vagaries of stone to discount the poorer material, for although they were not so concerned with the aesthetics of stone, they were most definitely concerned with the robust quality of the blocks extracted at local quarries. Evidence of prudent selection of red sandstone can be seen in the plain but serviceable railway bridges spanning the routes of the Shrewsbury and Chester and Shrewsbury and Crewe lines where they run across the red sandstone outcrops of North Shropshire.

The New Red Sandstone of the combined Triassic and Permo-Triassic Systems dominates the landscape of the North Shropshire Plain. It is manifest in the hills which punctuate this lowland tract and is conspicuous in the abrupt sandstone cliffs which break through in the Nesscliffe and Hawkstone districts. It is, however, the quarries and the buildings associated with them that expose the stone to closer scrutiny. Exploited for centuries by architects, builders and masons of varying degrees of expertise and aesthetic sensibility, the sandstones of this district — excluding Grinshill stone which must be regarded as a special case — have featured on a diversity of structures within the neighbourhoods of the quarries and in the county town of Shrewsbury. Just a few of these sandstones have extended their marketing range to neighbouring regions, mostly in the secondary role as dressing stone to locally quarried stones. For the most part the red sandstones were confined to the buildings of their immediate localities of outcrop and it is here that their impact on the local architecture may be judged.

4
The New Red Sandstone of East Shropshire

The red sandstones, so prominent in the hills of North Shropshire, also outcrop along the eastern side of the county and in the adjacent districts of Staffordshire where the quarries at Penkridge and Codsall were important sources of building stone for the industrial towns of the West Midlands in the nineteenth century.

On the Shropshire side of the boundary, quarrying took place at several locations in the vicinity of Market Drayton, Newport and a number of villages immediately to the west so that many vernacular and ecclesiastical buildings were built of the red sandstones. The better building stones were once again found amongst the Lower Keuper rocks of Triassic age but in this part of Shropshire the Bunter sandstones — which used to be classed as Triassic rocks but are now ascribed to the Permo-Triassic — were frequently used as building stone, often with rather disastrous results. These Bunter sandstones are commonly friable and pebbly, especially in the Bridgnorth area to the south.

The Lower Keuper sandstones, which are concentrated to the east of the Bunter outcrops and astride the Shropshire-Staffordshire boundary, were quarried in Shropshire in the Claverley, Worfield and Tong districts. Once again the variations of colour noted in the previous chapter could be found with the paler 'white' sandstone constituting the most prized building material. Tong church, one of the most elaborate and decorative ecclesiastical buildings in Shropshire, is constructed of this superior stone.

The churches of this eastern side of Shropshire are, with few exceptions, the largest and most stylish in the county since this was the most progressive area economically and the churches were the most richly endowed. Nevertheless, several of the churches are amongst the most severely weathered buildings to be found in Shropshire simply because they were constructed of sandstones of indifferent quality, extracted from the New Red Sandstone outcrops. Those built at an early date have been the subject of repeated restoration through the centuries. This is especially true of the ecclesiastical buildings of the Bridgnorth district where the Bunter sandstones of the Permo-Triassic System are concentrated.

During Permian times this midland region of Britain was probably located even nearer to the equator than it was in the subsequent Triassic times when, as previously described, the Grinshill sandstone was being formed. Being some eight to twenty degrees north of the equator the aridity would have been pronounced, so that although some of the sands of the Bridgnorth area settled out in lake water, the majority accumulated in terrestrial dunes much blown about by shifting winds. The resulting Dune Sandstone is cross-stratified, the sloping partings in the rock being very apparent in roadside cuttings and quarry faces (Fig. 4).

The rounded quartz grains denote the aeolian origin of the sands and wherever these grains are bound by unstable minerals the resulting sandstone is most likely to be friable and easily worn away. In the extremely dry oxidising environment, most of the desert sand grains were individually coated with iron oxide, hence the bright red colour of the stones as they appear today. In some localities the pore spaces between the grains were so open that solutions could filter through and chemically rot the stone from the inside. None of the sandstones could be relied upon as building material, but despite this they were in common use in the Bridgnorth district for several centuries.

In addition to the friable nature of the Bunter sandstones, there were a number of other problems. The cross-stratification of the dune structures was not only visually disturbing but it effectively weakened the individual stones since the partings accelerated the seepage of damaging solutions into the sandstone. Grooves were etched into the exposed faces of the stones. These can be easily identified on some of the masonry fabric of St. Leonard, Bridgnorth, and St. Mary Magdalene, Quatford.

Other problems bedevil the Bunter sandstones which include the Pebble Beds and the Lower and Upper Mottled Sandstones. As the term implies, the Mottled Sandstones are not of uniform colour for although predominantly red they contain streaks, lenses and patches of yellow material resulting from the differential concentrations of minerals. Generally, these sandstones are soft and easily weathered and therefore of more use as building sand rather than stone. Where they are sufficiently coherent for stone construction they have been used in cottages and farm outbuildings, for example in the district to the north of Sheriffhales.

Apart from the conglomerate which lies at the base of the Lower Mottled Sandstone, the Pebble Beds are the most unevenly textured of the Bunter sandstones. Nevertheless, they have been very widely used as building stone. The pebbles are not only of differing sizes (less than one inch to as much as nine inches), but also of varying colour. This is due to the fact that the pebbles were derived from a variety of rocks eroded from the Mercian Uplands, which in Permian times lay to the south of the midland deserts. Rainstorms generated over these uplands caused swiftly-flowing streams to cascade on to the desert flats to the north. With the abrupt loss of stream velocity on the dry plains the pebbles, sands and muds were dumped *en masse*, especially at the ends of the valleys or on the shores of temporary lakes where the ill-sorted accumulations assumed the form of spreading fans (Fig. 6). These consolidated into the Pebble Beds — not the most promising of building stone material. However, despite their somewhat inauspicious origins, the Pebble Beds vary sufficiently to offer some scope for builders and engineers. The toughest were selected by railway engineers who were not too concerned with the appearance of the stone, but some were used by church architects with rather more success than might be thought possible.

The key to the potential uses of the Pebble Bed sandstones lies within their extreme variability. Obviously, if the pebble content dominates, either in terms of the dimensions of the pebbles or in their frequency, then the resulting rock will be far too coarse and unevenly-textured for any building purpose other than the construction of crude walling. If, on the other hand, the pebbles are inconspicuous, the rock will be suitable for a range of construction purposes, providing also that the sandstone matrix in which the pebbles are embedded is resistant to weathering.

If selected with care and an awareness of their shortcomings, the Bunter rocks of the Permo-Triassic System can serve several useful roles in the construction industry. As building stones, their lack of uniformity must always make them suspect. They vary considerably in reliability even within a single quarry — a legacy of their mode of formation — so meticulous inspection of each stone is essential. Despite the problems, much of the vernacular architecture, quite a substantial amount of ecclesiastical architecture and a number of engineering projects depended on supplies of stone from the many small sandstone quarries scattered over the Bunter outcrops. A few examples should serve to illustrate the varying attributes of these stones and highlight the deficiencies that have subsequently led to problems with the masonry fabric of churches in particular.

The church of St. Mary Magdalene, Quatford, is perched on an outcrop of Lower Mottled Sandstone where it rises to Camp Hill east of the Severn and to the south of Bridgnorth. The Bridgnorth to Kidderminster road slices through the sandstone exposing the cross-stratified structures of sand dunes formed during the Permian Period. The rock is bright red in colour and is extremely friable. It was used for much of the church fabric with unfortunate results. Most of the sandstone structure dates back to the early eighteenth century when the tower and nave were rebuilt, but the south aisle and the south porch were constructed as late as the mid-nineteenth century. Regardless of date, much of the red sandstone has weathered badly, not surprisingly in view of the nature of the rock and the very exposed location of the church. Recent renovation has replaced the rotted east window and refurbished the tower masonry, but there is still evidence of decay on the south porch where the red sandstone blocks have worn back so much that the mortar stands proud of the faces of the stones. Here and there a tougher pink sandstone accentuates the inferior nature of the local bright red stone. This is Alveley stone, which was used as dressings to the south aisle wall and on the porch during a major restoration by a local architect, R Griffiths, in 1857. Alveley stone is not Permo-Triassic. It derives from Carboniferous sedimentation and being a good quality stone it was exploited as one of Shropshire's main commercial stones until well into the twentieth century.

Another very different kind of rock can be identified in the external masonry of Quatford church, especially in the chancel which dates back to Norman times. Despite appearances, this rock has weathered well, much better in fact than the sandstones used for the rebuilding work of the nineteenth century. The rock is tufa which, being full of cavities and rough-textured, tends to resemble a fossilised sponge. Though it appears frail, it is essentially a very tough stone, being a chemical limestone deposited by streams issuing from beneath limestone outcrops. Such streams are invariably

The porch of St. Mary Magdalene, Quatford, built 1857 to the design of a local architect, R Griffiths.

over-charged with calcium bicarbonate dissolved out of the limestone rock as the water threads its way along the joints and bedding planes. When the streams emerge into the atmosphere from their subterranean channels they deposit lime-saturated water on any object over which they flow. The moisture evaporates to leave a coating of lime on twigs, leaves and grasses. This 'frosting' is soft initially but soon hardens with exposure to the atmosphere. The entrapped vegetation is thus petrified and converted into a cellular but very serviceable stone. Rarer types of tufa are streaked with impurities which cause colour variations in the stone. They can be polished to simulate marble for purely decorative purposes.

Quatford church may be cited as an extreme example of a church with a patchwork of walling stones. Obviously, there has been little attempt to keep to a uniform stone throughout successive phases of rebuilding. Other churches in the red sandstone districts of eastern Shropshire display more muted contrasts in their stone fabric as nineteenth-century restorers availed themselves of quality pink and red sandstones obtained from commercial quarries at Grinshill, Highley and Alveley in Shropshire; Codsall, Hollington, Penkridge in Staffordshire; Mansfield in Nottinghamshire and Runcorn in

Tufa and red sandstone juxtaposed, emphasising the weathering of the less resistant sandstone — St. Mary Magdalene, Quatford.

Cheshire. Efforts were made to match the stones as closely as possible to the existing masonry but there is usually a tell-tale discrepancy between the older, weathered stonework and the pristine and mechanically precise masonry of more recent date. In addition to having the advantage of a later installation the commercial freestones are inherently tougher.

Nineteenth-century aesthetes such as Augustus Welby Pugin and John Ruskin deplored insensitive restorations which only achieved discordant effects in the masonry fabric of ancient buildings, but there was a dilemma to resolve in all areas where the local stones were of inferior quality. The choice between the two extremes of indulging in a *laissez-faire* romantic appreciation of mouldering decay and of tackling the problem of weathered masonry led to some bitter arguments during the nineteenth century with artists, architects, masons and builders assuming very different attitudes on the vexed question of restoration. During church restorations masons, quarry owners and building committees were often caught up in the fray and the incumbent was sometimes called upon to act as mediator. Architects persuaded of the expediency of using the local stone for maintaining continuity with existing church fabric and for keeping down costs simply risked a repetition of the mistakes of their predecessors and failed to resolve the basic problem of poor quality stone. Others who ventured to introduce alternative stones were often castigated for their insensitivity.

The church of St. Leonard, Bridgnorth, suffered damage during the Civil War but the more insidious centuries-long weathering of the sandstone fabric was the main cause of its dangerous condition in the middle of the nineteenth century. This impelled the Rev G Bellet to apply to the Incorporated Church Building Society, in June 1860, for financial assistance with the urgent and substantial task of rebuilding. The chancel had already been restored in 1846 with the aid of the architect F J Francis. In 1860 a London-based architect, William Slater, was engaged to undertake the more extensive repairs necessary on the nave and the aisles which involved internal as well as outside masonry work. This was phased over ten years and entailed the use of local red sandstone, some of which was offered by T C Whitmore, Esq from quarries on his lands to the north of Bridgnorth. These were located on the High Rock outcrop of Bunter sandstones.

In 1870 the main focus of concern became the tower, which Slater regarded as particularly dangerous. After much debate and correspondence, the building committee selected Town Mills stone for the new facing of the tower, having rejected Alveley stone on the grounds of cost. Thus the tougher and more reliable Alveley stone was turned down in favour of yet more Bunter sandstone. The Gloucester builder, Albert Estcourt, doubted the wisdom of the choice at the outset. When it was decided to raise the height of the tower in 1872 he refused to assume any responsibility for over-weighting the masonry. In the event his misgivings were justified, for even as early as 1873 some ashlared Town Mills stones were having to be replaced.

Prior to recent renovation of its sandstone masonry, St. Leonard's provided indisputable evidence of the adverse effects of using the local Bridgnorth stones which, as previously noted, derive from the Bunter rocks of the

Spalling sandstone blocks in the walling of St. Leonard, Bridgnorth.

Permo-Triassic System. Many stones were spalling so badly that the initial ashlared blocks were defaced. In addition, the cross-stratified structure of many of the stones caused deep grooves to form as weathering progressed. The determination of the building committee to hold down costs must have overruled the builder's suspicions of the local stone, managing at the same time to persuade the architect to use it, but Slater's partner, R H Carpenter, offered some moral support to Estcourt when replacement stones were needed in 1873. However, the two came into conflict in 1882 when a pinnacle, constructed in this case of Alveley stone, was brought down in violent gale. Defective dowelling and the bad selection of stone at Mr Lane's quarry at Alveley were cited as reasons for the failure of the pinnacle, but the builder protested. He not unreasonably disclaimed any responsibility for the violence of the storm and for a decade of weathering on the stone. Twenty years of controversy and haggling over costs and liability passed in this restoration work, during which time both William Slater and Oliver Estcourt, one of the brothers in the Gloucester building firm, died. A succession of three vicars was also involved, so there must have been several people much exercised by the troublesome red sandstone of the district.

The selection of Alveley stone for the pinnacles would seem to have been judicious, for this was a proven stone, resistant to weathering, whereas the relatively soft Bridgnorth stone could hardly be expected to suffice in such an exposed position. The problem arose not from the calibre of the stone but from the difficulty of procuring an adequate depth of bed of uniform stone from which to carve each pinnacle. Carpenter's Clerk of Works recommended the use of several securely dowelled stones to make up the required height of the replacement pinnacle in preference to the use of deeper but potentially less consistent stones.

Carpenter's problems with the pinnacles at St. Leonard's coincided with restoration work on St. Mary's, Market Drayton. Once again the local red sandstone was the fundamental cause of the poor state of the masonry, especially on the aisles. Elsewhere on the church, a tougher grey sandstone had been used and needed less repair. R H Carpenter, the architect appointed by the building committee, was assisted in this case by his new partner, W B Ingelow. George Fletcher, the very same Clerk of Works who was concerned with St. Leonard's, also supervised the restoration of this other red sandstone church. The bulk of the new masonry needed for the substantial rebuilding of the aisles, the construction of a porch and for refurbishment of the nave and chapels was executed in Hinchley Mill stone which was quarried on the Bunter Pebble Bed Sandstone outcrops about one kilometre east of Market Drayton. Once again, a Carboniferous sandstone, on this occasion quarried at Old Springs just across the border in Staffordshire, was turned down in favour of the Bunter sandstone which was tendered at a lower price. Despite this the restoration was expensive, costing over £7,700 on completion in 1889.

Most of the red sandstone used has weathered better than the Bridgnorth sandstones but there is still considerable wear to be seen on the late-nineteenth century masonry. All Pebble Bed sandstones are prone to differential weathering of the sandstone matrix and the pebbles. As the latter are commonly of tough quartz they stand proud of the faces of the sandstone blocks until a point comes at which they fall out of the stone and leave disfiguring cavities. Clayey lenses in the rock also weather relatively quickly, as do the cross-stratifications. In time the smooth ashlared faces of the stones become grooved and pitted. The Hinchley Mill stone used on St. Mary's shows all these blemishes.

The very meticulous restoration of the tower at St. Mary's involved the re-use of sound stones which were carefully removed, numbered and re-set in the replacement masonry. Stones from Grinshill and Runcorn were also incorporated into the tower walling and best Idlestone, a Carboniferous sandstone from Yorkshire, was used for the paving and steps because of its resistance to attrition and its non-slip quality. Runcorn stone was combined with alabaster on the low screen wall of the chancel and polished Hopton Wood stone — a very crystalline Carboniferous limestone from Derbyshire — was used for the altar steps. All these were, of course, proven stones and were marketed throughout the country. The same could not be said of the Pebble Bed sandstones which only served a localized market. However, throughout

Rough-hewn Pebble Bed Sandstone from the local Water's Upton quarry used for bridge construction on the Wellington and Market Drayton Railway.

the New Red Sandstone districts of north and east Shropshire the Pebble Bed sandstones proved to be extremely useful in the construction and renovation of many churches, as well as for the building of a wide variety of domestic and agricultural premises.

As engineering stones, the Pebble Bed sandstones had much to recommend them, for they could be quite robust providing the sandstone matrix was cohesive. The rough texture was of no consequence in the building of sturdy railway bridges, for there was rarely any attempt to cut and dress the blocks smoothly. The village quarry at Water's Upton provided stone for the construction of nearby bridges over the Wellington and Market Drayton Railway in the 1860s. The rock-faced and rough-textured stones were cut to various sizes and built into the random-coursed masonry. The canal builders also used Pebble Bed sandstones to good effect in the construction of the aqueduct that carried the Shrewsbury Canal over the swampy tract of land known as The Weald Moors to the south of Kynnersley. The tough and rather gritty stone was also used in the church of St. Chad at Kynnersley and on a number of cottages and farm buildings close to the quarry at Wall.

During the nineteenth century it became quite acceptable for church architects to design a rural church with rock-faced and random-coursed walling masonry and possibly, though less commonly, with rubblestone masonry. Both types of construction were compatible with rural settings and were used by Benjamin Ferrey in the building of two red sandstone churches in eastern Shropshire during the 1860s. At this time influential architects were attempting a more sensitive use of local materials in relation to the environment.

At Sambrook to the north-west of Newport, Ferrey made excellent use of the local Pebble Bed sandstones for the irregular rubblestone walling of St. Luke's church. His sources of stone were the quarry at Sambrook, which yielded a dark red medium-grained sandstone, and quarries at Standford Bridge and Calvington where similar but coarser-grained sandstones were available. The rough sandstones were perfectly adequate for the general walling but better quality freestone was essential for the dressings. This was quarried in Lower Keuper Sandstone outcrops at Sutton and Chatwell to the south-east of the building site and at Betton to the north.

Ferrey adopted a more sophisticated style of ecclesiastical architecture for St. Michael, Chetwynd, choosing a more refined masonry technique for the more ambitious and imposing structure. The materials were basically the same however, and included locally quarried Pebble Bed sandstones and the

Irregular rubblestone walling combined with dressings of finer sandstones at St. Luke, Sambrook.

finer freestones from Sutton and Chatwell. Ferrey wisely refrained from using an ashlared finish on the walling stones, preferring to keep them roughly rock-faced for composing an interestingly random-coursed outer face of different-sized squares and rectangles. This rough-textured walling is set off by bands and dressings of the smoothly cut freestones. In the building of these two churches at Sambrook and Chetwynd, Ferrey had exploited the potential of the local stones for expressing the individual character of each design.

The poor building stone potential of some of the red sandstone rocks of the Permo-Triassic and Triassic systems is most easily recognized on churches of antiquity, partly because of the prolonged exposure of the stones to the atmosphere but also because of the long-established tradition of ashlaring any stone that was relatively easy to work. This practice intensified the vulnerability of such tender stone to the rigours of weathering. For centuries church builders had persisted in using local stone even when its weaknesses were difficult to ignore, largely because of the transport problems involved in the introduction of better material from distant quarries. Repeated patching up of the decayed fabric became an integral part of the history of a church.

This is certainly true of the medieval church of St. Nicholas, Newport, which suffered from several futile attempts at restoration with the local red sandstones down the centuries prior to the late nineteenth-century restoration by the architect J Norton. In defiance of tradition he used the hard red Codsall stone of Staffordshire for his major repair and re-building programmes of the 1880s and 1890s. The choice of Codsall stone has certainly been justified in terms of durability but some purists might consider the restoration work unsympathetic to the ancient fabric, especially that of the tower which dates from the fourteenth century, with repairs during the second decade of the seventeenth century when new battlements were added.

Sufficient space has been devoted to the New Red Sandstone outcrops of both east and north Shropshire to dispel any lingering misconceptions regarding the uniformity of the stones. Closer inspection has revealed a great diversity of colour, texture, and potential durability which can only be explained by the geological conditions responsible for the formation of the rocks.

When the extremely wide-ranging talents and skills of the persons involved in the use of the red sandstones are also taken into consideration, it is easy to appreciate why such a variety of architecture has resulted. Nationally renowned architects and local builders used the sandstones in many different ways, but their capacity to assess the calibre of the stones in relation to specific purposes was a crucial factor in the success of their endeavours. The function of a proposed building and the sophistication of its design should always be leading considerations in the choice of stone. Given informed selection of stone at the quarries, the various sandstones of the New Red Sandstone districts have served the building industry tolerably well and have certainly made a colourful contribution to the architecture of the county.

5
The Old Red Sandstone of South Shropshire

The Old Red Sandstone rocks of South Shropshire, like the new Red Sandstone rocks already described, have been so named because of the occurrence of reddish tinges in some of the rocks. This colouring is associated with hot, dry oxidizing environments where plant life is negligible. Desert conditions, which dominated New Red Sandstone times, were probably not so pervasive in Shropshire during the Old Red Sandstone Period but plant life at this early date was so primitive that the effects were similar. However, the rocks of both geological systems are not uniformly red.

The all-embracing term 'sandstone' is especially inappropriate when applied to the rocks of the Old Red Sandstone System because the sandstones are everywhere interbedded with silts and limestones for reasons discussed below. Only the terms 'Old' and 'New' have any true relevance, being the simplest way to distinguish two major phases of continental sedimentation, the 'Old' occurring some 395-345 million years ago, the 'New' some 280-190 million years ago. During the intervening Carboniferous Period the flooding in of extensive seas and the subsequent establishment of dense tropical forests on delatic swamps were major events which resulted in the evolution of the Carboniferous Limestone, Millstone Grit and Coal Measure formations which were destined to be of such enormous economic significance in South Wales and Northern England. Shropshire's legacy of Carboniferous rocks will be discussed in the following chapter. The current chaper deals with the rocks of the Old Red Sandstone System but it is useful to note in passing that during the whole prolonged time-span between the start of the Old Red Sandstone and the close of the New Red Sandstone, part of the earth's crust which was destined to become the British Isles was located within about thirty degrees of the equator and therefore experiencing tropical and sub-tropical conditions.

The difference between the Old and the New Red Sandstone is not simply a matter of age. There is evidence of much more pronounced aridity in New

Red Sandstone times, especially during the Permo-Triassic Period. In the Midlands, the deserts were widespread and the accumulations of sands were deep. During the Old Red Sandstone Period much of the region which now constitutes South Shropshire and Herefordshire was located in an intermediate position between land and sea. The sea spread across much of southern England and resulted in marine sedimentation rather than the continental sedimentation which prevailed further north. In fact the Old Red Sandstone is termed Devonian wherever marine deposits occur, since the most typical area for such sediments is Devon. The landmass to the north was eroded under arid conditions and contributed vast quantities of detritus which settled in the sea to the south.

The southern part of Shropshire received some of this sediment which accumulated in river basins, lakes, estuaries and lagoons. Because of the fluctuating conditions between land and sea, the sediments were anything but uniform and comprised different sandstones, shales, limestones and conglomerates. In terms of their subsequent exploitation as building stones, the sandstones lack the deep beds of consistent material which can be found amongst the New Red Sandstone rocks. There is a paucity of good freestones. Another problem is the lateral replacement of the sandstone beds by other types of rocks so that the sandstones give out both vertically and horizontally and the quarries are small and scattered. There are no large commercial quarries comparable to those of the Lower Keuper building stone beds in North and East Shropshire. Most of the sandstone quarries of South Shropshire served a localized market.

Before describing some specific sandstones from the Old Red Sandstone System, it is necessary to draw attention to the fact that the Downton Castle Sandstone and the Holdgate Sandstone used to be classed as Old Red Sandstone rocks but are now considered to be of Upper Silurian age. However they still feature on Geological Survey maps as Old Red Sandstone outcrops and will therefore be included as such. As they are found on the western flanks of the Clee Hills, in Corvedale and also to the west of Ludlow they may be conveniently discussed along with other sandstones of South Shropshire.

The Old Red Sandstone rocks of Shropshire are concentrated in the Clee Hills. They include the Farlow Sandstone which was quarried for building stone during the nineteenth century for the construction of a number of rural properties and churches despite its dubious quality. Most of it is a current-bedded, medium-textured, yellow sandstone containing a scattering of quartz pebbles. It lies beneath younger Carboniferous rocks on the north-eastern side of Titterstone Clee Hill in the vicinity of Farlow Bank and Cleeton-St.-Mary. Some of the sandstone is rather soft and friable and is therefore indifferent building material, but there are some beds of tougher rock and several that provided 'firestones' suitable for use as hearthstones and fireplaces.

St. Giles, Farlow, was rebuilt in local sandstone to an Early English design by R Griffiths in 1858. Stones from the Norman doorway of the old church, which had been erected at the base of the hill, were hauled up Farlow Bank to

Quoins with diagonal tooling and drafted margins at St. Giles, Farlow.

the site of the new church which was built of the yellow sandstone quarried nearby. The new stone blocks were roughly squared and tooled and the stones of the buttress quoins were given a more elaborate treatment with diagonal tooling and drafted margins. All these textured stones obscure the deficiences of the Farlow Sandstone but around the windows flaws are conspicuous in the friable lenses of soft sandstone and in the hard quartz pebbles which produce a very uneven surface on stones that were intended to be smoothly dressed. The Geological Survey describes the Farlow Sandstone as being easily weathered, a fact confirmed by the present state of St. Giles.

At Cleeton-St.-Mary, the church designed by the Hereford diocesan architect, T Nicholson, was also in the Early English style and built of locally quarried Farlow Sandstone some twenty years later than St. Giles. The yellow sandstone is here rather soft and contains bands of quartz pebbles. The walling stones are now spalling badly and some of the rectangular blocks have rotted on their faces to depths of more than two centimetres. The dressing stone is Cotswold limestone quarried in the Painswick district. This 'Bath stone', a fine-grained limestone of Jurassic age, was used by church architects throughout Britain in the latter half of the nineteenth century as it was especially useful for window tracery. As railway transport facilitated the movement of stone around the country it became quite common for local stones to be dressed with superior commercial stones quarried at considerable distances from building sites. The Jurassic limestones were particularly suitable for moulding and carving and were in great demand for ecclesiastical buildings.

The Farlow Sandstone was just one of the Old Red Sandstone rocks used as building material in the Clee Hills. A similar buff and yellow sandstone from the older Clee Group sandstones was used in the construction of St. Lawrence, Burwarton, which was an estate church donated by Lord Boyne of Burwarton Hall. Both the church and the mansion were designed by an eminent nineteenth-century architect, Anthony Salvin, so it is surprising that such an unreliable stone was chosen for the external walling masonry of the church. The stone was quarried on Burf Hill, on estate land, so this choice must have been guided by expediency and possibly by some need for economy since inside little expense was spared. There was a lavish use of Purbeck 'marble' on the columns of the chancel and organ chamber, and the building firm selected was George Smith & Co. of London. In the event the use of local stone proved to be a false economy.

A panel of architects under the direction of Healing and Overbury of Cheltenham inspected the fabric of St. Lawrence for a series of Quinquennial Reports in the 1950s and 1960s, some eighty years after the church was built. Their comments are indicative of the poor calibre of the stone which was said to be face-shaling and laminating on the walling masonry and actually perished on the coping, string courses and gargoyles. By the 1980s the rock-faced walling stones were spalling badly in many places and the moulded dressed stonework of the west door was deeply weathered.

Fortunately, not all the sandstones derived from the Old Red Sandstone outcrops of South Shropshire are as unreliable as those just described. Several are very serviceable and have lasted well on buildings of much greater antiquity than the three nineteenth-century churches referred to. Two churches dating back to the Norman period incorporate stones from the older Ditton Series of the Old Red Sandstone which laps around the Clee Hills at a lower altitude than the younger Clee Group and Farlow Sandstone outcrops. The Ditton Series includes a variety of rocks other than sandstones and there is little uniformity amongst the sandstones themselves. Briefly, these may be described as reddish-brown in colour but there are some greenish-grey beds. Most are reasonably fine-grained sandstones and though there are some thicker beds, the majority are shallow and may be termed flaggy. There is nothing here to equal the deep freestone beds of Grinshll, so ecclesiastical and vernacular buildings are usually constructed of small stones and not uncommonly of rubblestone.

The two churches selected to illustrate the use of Ditton Series sandstones are located on the eastern side of the Clee Hills. They both incorporate a variety of buff, orange and purple-tinted sandstones which were quarried locally. Holy Trinity, Sidbury, was rebuilt by the Nevett Bros. of Ironbridge to designs of R Griffiths in 1881 from stone quarried close to the church and presented by C R E Cresswell of Sidbury Hall. This stone was matched in part to the refurbished Norman herringbone masonry and has weathered much better than the sandstones used on the three churches previously mentioned. In fact, the most severely weathered part of this church is the early eighteenth-century Cresswell chapel which, in conformity with the contemporary fashion for smooth ashlared walling, was constructed not of the local sandstone like the rest of the church but of a more easily cut greenish-grey siltstone. The temptation to choose a stone that could be obtained in bigger and more easily dressed blocks should have been resisted.

The Nevett brothers had earlier been engaged on the restoration of St. Mary, Stottesdon, in the 1860s, again using local stone, this time from a quarry on the glebe land. This has worn reasonably well in common with the stones of the Norman and Perpendicular masonry. The basic construction problem at this church is not so much due to the calibre of the stones chosen as to the injudicious siting of the church on soft marl foundations which has caused the building to settle to the north, thereby endangering both the external and internal masonry.

Geology is implicated in the durability of a stone building for several reasons. Obviously the calibre of the stone quarried for construction purposes is of prime importance, but the type of natural foundation beneath the structure is just as crucial to the survival of the building. Building and masonry techniques are also of vital importance, for bad workmanship on good stone can be detrimental to the structure. An understanding of the vagaries of stone is imperative if the construction is to last for at least as long as the natural life-span of the stone itself. Often poor selection of mortar has triggered the chemical decay of stone and the use of iron cramps has accelerated decay. Two churches built of Old Red Sandstone materials and located to the west of the Clee Hills usefully illustrate these various hazards.

The first is St. Mary, Bromfield, which is situated close to the confluence of

St. Laurence, Ludlow — weathered siltstone showing advanced weathering along the mortar courses.

the Onny and Teme, approximately three-and-a-half kilometres to the north-west of Ludlow. St. Mary's church, which dates back to the Norman period but incorporates later phases of construction, was in desperate need of restoration by 1889 when C Hogson Fowler of Durham reported on its condition to Lord Windsor of Oakly Park. The architect stressed the very dangerous state of the tower walls which he found to be built of small stones set in poor quality mortar and therefore not sufficiently cohesive to withstand the weight and vibration of the bells. Work was also needed on the north wall and, mindful of the desirability of matching the new with the best of the early masonry work, especially that on the south side of the nave, the architect specified the use of stone from the nearby Felton Quarry. This was situated approximately two kilometres to the east of the church where the River Corve cuts through the Ledbury Beds of the Old Red Sandstone. Here, pink and purple 'red sandstones' were quarried for use in the local district, especially in Bromfield village and in Ludlow. Fowler used the Felton stone for the rubble backing and also for the facing stones which were only slightly dressed and laid to courses. This masonry technique was well suited to the rusticity of the stone which was micaceous and occasionally flaggy and coarse in texture. Most of the vernacular buildings in the locality are of coursed rubblestone.

The Felton stone along with the clean sharp sand, which was obtained from the river bed close by, were the only local materials used for the restoration. The architect chose Westmorland slates for re-roofing the chancel, Runcorn stone for the dressed stone work and York and Mansfield stones for the new paving and steps inside the church. The Old Red Sandstone was evidently considered suitable only for the general walling.

Stone from Felton Quarry had also been used in various stages of reconstruction of the second church, St. Laurence, Ludlow. This fact came to light when Arthur Blomfield suggested the use of Runcorn stone for the restoration of the tower in 1889. The building comittee was adamant that the stone selected should be the same as that of the original masonry and in searching through the Churchwardens' Accounts of 1467-71 verified this as stone from Felton Quarry. A neighbouring quarry on the Ledbury Beds of the Old Red Sandstone outcrops was opened up to provide stone for the restoration.

The poor state of the tower in the late nineteenth century was due to several factors, including the indifferent technique of the fifteenth-century masons. The foundations had been badly laid with an assortment of stones of different sizes and shapes which had been loosely assembled and covered with large slabs. These failed to guarantee any stability to the tall tower so that signs of movement materialized during the first century of the tower's existence as the west side began to sink and cracks opened up in the walling masonry. These were bridged with iron cramps — a temporary expedient and no solution to the fundamental problem.

The walling itself was deficient, having been assembled of two facings of ashlared stonework between which rubble stones had been only loosely thrown in. The walling lacked cohesion and needed reinforcing. This was achieved by the careful bonding in of tough York stones wherever cracks had

Renovated Norman herringbone masonry in an assortment of local sandstones at Holy Trinity, Sidbury.

Ashlared siltstones on the Cresswell Chapel of Holy Trinity, Sidbury.

The coarse-textured, purple-tinted Holdgate Sandstone containing fragments and pebbles worn from pre-existing rocks.

A current-bedded Keele Bed sandstone block in a wall of Telford's St. Mary Magdalene, Bridgnorth.

The south wall of St. Mary Magdalene, Bridgnorth.

The south front of St. Margaret, Ratlinghope, built of Stiperstones Quartzite.

The porch of St. Peter, Diddlebury, with red sandstone dressings to siltstone walling masonry.

Detail of the weathered siltstones on St. Peter, Diddlebury.

appeared and by the use of slate dovetail cramps which, unlike iron cramps, have no adverse chemical reaction to the stones. In addition, cement was poured into the core to fill the gaps between the rubble stones.

The old facing stones which had peeled at the surface had to be replaced by new stones in the walling. Even more drastic replacement of stone was needed on the windows. These were in a very dilapidated state since they had been carved out of local olive-green siltstones which lacked resistance to weathering. Siltstone is what is termed a laminated stone, being formed out of thin layers of mud which have become compacted and indurated into rock — a type of rock that splits and peels fairly easily. Ease of carving may have tempted the medieval masons into using the soft local siltstone, but there was no excuse for ignoring one of the basic principles of sound masonry which is to lay the stone on its natural bed. In placing the siltstones with the laminations and bedding vertical the early masons had encouraged the rapid weathering of the window stones since rainwater quickly converted the partings into grooves. Even when siltstone is laid on its true bed, there is no guarantee of good weathering potential, for it is usually very clayey and soft. The siltstones beneath the string course at the base of the south transept wall of St. Laurence are now very decayed, especially along the mortar courses where seeping water has accelerated chemical weathering.

St. Laurence, in many respects a fine specimen of medieval artistry and craftmanship, is wanting in terms of constructional technique with the result that successive generations of masons have been engaged in repairing damage

St. Laurence, Ludlow — spalling sandstone. Weathering accentuates the bedding structures in the cut stone.

Neatly coursed, rock-faced masonry of Onibury stone contrasts with the ashlared red sandstone blocks of the buttress quoins at St. Cuthbert, Clungunford.

largely attributable to the mistakes of their predecessors. However, the early masons cannot be held totally responsible for the wear of the stone and in any case their options were limited by restricted transport facilities. Blomfield, with railway transport at his service, was able to select Hazelstrine stone from Staffordshire when needing a quality freestone for the carved figures of the saints placed in niches in the upper windows. However, in deferring to the Building Committee over the use of Felton stone for the walling, he merely repeated some of the errors of the early masons. Some of the ashlared stones are spalling badly and are quite decayed in places. In general, the local sandstone is better suited to the construction of rubblestone premises.

Two sandstones that dominate local stone buildings in their districts of outcrop are the Downton Castle Sandstone and the Holdgate Sandstone which, as already explained, appear on Sheet 166 of the Geological Survey as Old Red Sandstone rocks. Both sandstones were probably deposited in intermediate locations between land and sea — environments suggested include lagoons, intertidal beaches and deltas where marine and continental conditions alternated in response to oscillations of sea level. Hence their recent inclusion in the Silurian Period which was predominantly marine. The Downton Castle Sandstone appears in the area west of Ludlow, in the vicinity of Ludlow and in Corvedale to the north-east; the Holdgate Sandstone outcrops along the eastern side of Corvedale. Both sandstones are interbedded with shales and marls and therefore impress as the best building stones of their respective districts. They are quite distinctive stones.

The Downton Castle Sandstone is recognizable by its yellow-buff colour. It is mostly fine-grained and micaceous and although it may be thinly bedded and even flaggy in some localities, it can also be found in thicker beds from which stones of reasonable size could be quarried. The presence of current-bedding in the rock implies deposition in shallow waters, possibly as beach sand. This sandstone has proved to be very useful as tilestone and freestone in South Shropshire and the adjacent parts of Herefordshire. Onibury Quarry provided building stone for the Ludlow-Onibury-Stokesay district and for several villages to the west. The tower of St. Cuthbert's, Clungunford, which was rebuilt in 1895, was constructed of rock-faced rectangular blocks of Onibury stone that were laid in courses of varying depth.

The Holdgate Sandstone features on rural buildings throughout Corvedale, especially in the neighbourhoods of Broncroft and Holdgate, where the distinctive purple-tinted 'red sandstone' is instantly recognizable. It is a coarse-textured rock and is believed to represent the infilling of channels that were cut through mud-flats by rivers draining from a tropical landmass and bringing down vast quantities of red-coloured detritus.

The buildings cited in this chapter and the building sandstones used in their construction are indicative of the very wide variations in the rock outcrops of the Old Red Sandstone System. The differences found amongst the sandstones of the New Red Sandstone System outcrops of both North and East Shropshire, though appreciable, have not resulted in such a diversity of architecture. In fact, red sandstone buildings typify these two areas. In South Shropshire, the variety of both the natural landscape and its associated architecture is remarkable and the Old Red Sandstone outcrops have contributed a great deal to this diversity. The remaining sandstones of the county have more restricted distributions but they are not devoid of impact on the architectural scene.

6
Carboniferous Sandstones

Any mention of Carboniferous sandstones immediately conjures up images of Pennine towns, villages and farms in Lancashire and Yorkshire where the Millstone Grit and Coal Measure sandstones feature prominently on a wide variety of municipal, ecclesiastical and domestic buildings. Shropshire's Carboniferous outcrops never gave rise to Pennine-type architecture or, for that matter, to the all-pervasive coalfield industries which characterize the northern counties, but they did feature on more premises in the county than might be readily appreciated.

There are five Carboniferous districts within Shropshire: in the north-west — Oswestry-Morda; the centre — Shrewsbury-Hanwood; the east — Wellington-Ironbridge-Oakengates; the south-east — Severn Valley-Wyre Forest; the south — on the upper slopes of the Clee Hills. It is in the first of these five that rocks of Millstone Grit age reached any significance as building stones. Elsewhere, the building sandstones were mostly derived from the younger Coal Measure outcrops.

In the Oswestry district the important building sandstone of Millstone Grit age is known as the Cefn-y-Fedw Sandstone, the outcrop occurring in a zone bordered by the Carboniferous Limestone in the west and by the Coal Measures in the east. All three zones are aligned north to south and continue across the River Dee into the Ruabon-Wrexham district where big commercial quarries developed in the nineteenth century to supply freestones for municipal buildings in Liverpool and the coastal resorts of North Wales.

The Millstone Grit sequence is believed to have evolved on deltaic lowlands prior to the establishment of the dense tropical forests that were later compacted into coal seams. These deltas extended at the expense of the Carboniferous sea which had gradually transgressed over the British Isles covering some, but not all, of the Old Red Sandstone rocks referred to in the previous chapter. This sea invaded some 345 million years ago and deposited marine sediment mostly in the form of limestone which was destined to be of economic significance as agricultural and industrial lime and as building stone during the eighteenth and nineteenth centuries. As the sea retreated the deltaic swamps became prime locations for the vast accumulations of grits and sands brought down from the Old Red Sandstone landmass. The swamp forests rooted on the deltas were periodically invaded by the sea which advanced and retreated in response to successive warpings of the earth's crust. The resulting Coal Measures, therefore, contain rocks other than coals, including shales, siltstones, sandstones and ironstones — a geological endowment of immense value to man. The Carboniferous Period probably ended approximately 280 million years ago and, during its span of some 65 million years, Britain wandered about in tropical latitudes, experiencing humid equatorial-type conditions in which the swamp forests flourished. With a northerly drift towards arid climates as the Carboniferous Period gave way to the Permian Period, Britain was on course for staging the New Red Sandstone desert environments described in previous chapters.

The Cefn-y-Fedw Sandstone of the Oswestry district is known as Sweeney Mountain stone, since the hill constitutes the most conspicuous outcrop of the sandstone and was the scene of some important quarrying activity in the eighteenth and nineteenth centuries. Proximity to the Llanymynech Branch of the Ellesmere Canal, which linked up with the Montgomeryshire Canal at Carreghofa in 1797 and eventually gave access to Newtown in 1821, provided the impetus for the marketing of the stone southwards along the Shropshire-Montgomeryshire border. This was boosted by the opening of the Oswestry and Newtown Railway in the 1860s. Previously, the sandstone had only been marketed in the town of Oswestry and in the surrounding rural districts where it proved extremely useful not only as a building freestone for a variety of premises and for rural bridges but also as rubblestone on farm buildings and walls. In addition it was a source of sand. Various items including corn bins, pig troughs, flags and gutterstones for stables and cattle stalls as well as cheese presses and stone sinks were supplied by the quarries. Prior to the factory production of such items from alternative raw materials, stone quarries were the main sources of supply.

The geological conditions of the Sweeney Mountain sandstone outcrop made this diverse production possible. There are three main divisions of the sandstone visible at the face of Tower Quarry, one of the main quarries on the hill. The topmost layers are extremely friable and easily weathered and therefore the source of much of the loose sand which has accumulated on the floor of the quarry. Beneath are beds of a more cohesive sandstone from which the massive freestone blocks were cut. According to an advertisement of 1821, these were available up to twenty-seven feet in length and were the main source of building freestones and engineering stones. Being fine-textured this sandstone was ideal for ashlar work and consequently featured on local mansions such as Sweeney Hall, Porkington and Aston Hall. Underneath the freestone beds, a rougher sandstone was quarried for rubblestone which was much in demand for local farm buildings in the eighteenth and nineteenth centuries. This is a type of quartzitic sandstone where silica is concentrated in veins and nodules so that the disfigured stone is entirely unsuitable for ashlar. Nevertheless, it is usually a robust and durable stone and ideal for rural premises. It was also used for gateposts and rough walling and its thinner beds for flagstones.

The Sweeney Mountain quarries located on Sweeney estate could be leased for working in the nineteenth century. Naturally, they were the main source of stone for Sweeney Hall, built for Thomas Netherton Parker in the early years of the century. This was a plain five-bay, two-storey house with tall Tuscan pilasters — a classical design which required a good quality freestone for architectural authenticity. The house itself was actually situated on the Bunter Lower Mottled Sandstone of the Permo-Triassic outcrops of North Shropshire but the estate lands extended westwards on to the Carboniferous rocks which include the Cefn-y-Fedw Sandstone. A Rock Account Book of 1817 signed by T N Parker shows just how much use was made of this resource. Apart from masonry work on the main house, it provided stone for the construction of a summer-house built for family use at the top of Sweeney Mountain. The rougher quartzitic sandstone proved extremely useful for repairs and extensions to Sweeney estate properties as building improvements became imperative in the nineteenth century when most landowners around Britain were endeavouring to make up for the neglect of previous centuries.

The Sweeney Mountain quarries were a source of income for Sir Baldwin Leighton of Loton Park in the nineteenth century as the Loton and Sweeney estates were linked by family ties. The quarries benefited from the construction of several new churches along the western borders of Shropshire. Sweeney freestone often competed with Grinshill, Shelvock and Cefn stones (the latter from the Ruabon district of Denbighshire) as a dressing stone to

Varying tones of Sweeney Mountain sandstones in the walling of St. John Evangelist, Pool Quay.

the rougher local stones of the Welsh borders. However, St. John Evangelist at Pool Quay was almost exclusively constructed of Sweeney Mountain stone to a design by the Shrewsbury architect, S Pountney Smith, in the 1860s. This was a large church erected at the expense of the Earl of Powys with Sir Baldwin Leighton granting the architect permission to obtain the stone free of charge for royalty. The stone was neatly squared and dressed and laid to level courses of varying depth.

Much of the stone from the Tower and Nant quarries on Sweeney Mountain was used for less ambitious construction and featured on many domestic premises and chapels in the immediate locality. Hopes that the Llanymynech Branch of the Ellesmere Canal would bring the stone into widespread use in the county of Montgomeryshire never reached fruition and the quarrying activity never became large-scale. The rockmen and masons employed at the quarries were invariably local men and often from the same few families connected with the Sweeney estate. Whenever there was a lull in quarrying they were kept busy with general labouring work. The stone industry was therefore essentially a local enterprise with some masons operating from workshops in nearby Oswestry. Without the canal and the railway, the market range of the stone would have been restricted to the immediate district. As it was, these lines of communication extended its market range as a dressing stone rather than as a major freestone, so that Sweeney stone fulfilled a role similar to that of Shelvock stone in the construction industry.

As an engineering stone, Sweeney freestone was incorporated into the fabric of some fine local bridges. The repairs to Llanyblodwel bridge in 1827 necessitated consignments of ashlar stone from Tower Quarry. Telford's Rhydmeredith bridge of 1810, with its eighty-foot wide single arch span, included dressed Sweeney Mountain freestone in the arch voussoirs. More modest local bridges were often constructed of rough but serviceable stones from the same quarries. The stones quarried on Sweeney Mountain were of considerable importance in the Oswestry district for a diversity of uses during the nineteenth century but were probably unknown in other areas except for the border tract south as far as Newtown.

Two other Shropshire sandstones of Carboniferous age gained much wider reputations during the eighteenth and nineteenth centuries. This was largely due to their undoubted quality, but a very important factor in their success was proximity to the Severn. This river was the major transport artery of eastern Shropshire and counties to the south prior to the opening of the Severn Valley Railway in 1862. Highley stone, a yellow-grey sandstone from the Upper Coal Measures, outcropped in massive beds in the vicinity of Highley and Stanley to the west of the Severn. It was a building freestone of ashlar quality suitable for external walling masonry and as such was one of the country's leading stones examined by the commissioners responsible for selecting stone for the new Houses of Parliament in 1839.

Bridgnorth Bridge, located upstream of the quarries and extremely vulnerable to flood and impact damage on this busy stretch of the Severn, was in frequent need of repair work on its masonry in the eighteenth and nineteenth centuries. For this, the Carboniferous Highley stone or similar stone from Arley to the south was used on the facework and the decorative elements such as the parapets, whereas the inferior Triassic red sandstone from High Rock was used for the infill.

As a durable and aesthetically pleasing stone, Highley stone featured on several bridges spanning the Severn including those at Bewdley, Stourport, Worcester and Gloucester. It appealed to bridge engineers for much the same reasons that Grinshill stone did further north. It was a durable stone, available in large dimension blocks, capable of taking an ashlared finish and ideal for carving. For the construction of Bewdley Bridge in 1798 Thomas Telford obtained Highley stone from Arley Quarry which was only four miles upstream of the building site.

Lavish use was made of stone in bridge construction during the eighteenth century but by the second half of the nineteenth century bridges were increasingly built of iron, brick, fabricated stone and concrete since these became progressively less expensive alternatives. Stourport bridge, for example, completed in 1870, incorporated brick and iron but Highley stone was still required for the abutments, land arches, parapet walls and all decorative elements such as the strings, cornices and the moulded caps of the pilasters.

Highley stone had a long and distinguished history of engineering use in the construction of impressive bridges spanning the Severn downstream of Bridgnorth but it was also of architectural significance. Its resistance to weathering combined with its suitability for intricate carving promoted continuous demand from ecclesiastical architects who, even when working primarily in other stones of the district, would specify the use of Highley stone for such vulnerable, exposed features as battlements and gargoyles.

Alveley stone was quarried on the opposite side of the Severn valley to Highley and from younger rocks of the Upper Coal Measures known as the Keele Beds. These were generally of a red colour, reflecting the increasing aridity of late Carboniferous times. The stone was often selected by church architects for the decorative elements of their new buildings and also for strengthening structures which had been built of less durable stone at a previous date. For example, the medieval church of All Saints at Claverley was largely constructed of sandstones quarried out of local Lower Keuper outcrops but Alveley stone was used for buttresses which were introduced later to ensure structural stability.

The use of Alveley stone as a building material reached into the adjacent county of Worcestershire. It was chosen for the external walling masonry of St. John's Church in Kidderminster which was rebuilt in 1892. Because of its attractive colour it had ornamental uses as well. The chimney pieces in the classrooms of the Parish Schools of St. Mary Magdalene in Worcester, built in 1883-4, were cut out of Alveley stone. It was also used externally on monuments such as the tower at Abberley. This was mostly constructed of local sandstone but Ham Hill stone from Somerset was used on the upper stages and Alveley stone was introduced for decorative emphasis on the strings and doorway below.

Grindstones were produced at Alveley Quarry from a tough and gritty rock which, like all the best sandstones, had a high silica content. Unfortunately, this had serious implications for the men working the stone and the high incidence of silicosis amongst them is cited as one of the main reasons for the closure of the quarry in the early twentieth century.

Keele Bed sandstones also outcrop on the western side of the Severn and extend northwards to connect with similar outcrops in the East Shropshire coalfield. When selecting stone for his new church of St. Mary Magdalene in Bridgnorth in the early 1790s, Thomas Telford wisely discounted the local New Red Sandstone of which St. Leonard's had been built and concentrated his search amongst the Carboniferous sandstones outcropping to the west and south of Bridgnorth. Records reveal repeated sorties by masons and building trustees, sometimes in the company of Telford himself, for the purpose of inspecting any promising sources of stone in the Canton Brook and Dunval

Smooth rustication on the west front of St. Mary Magdalene, Bridgnorth.

The tower of St. Mary Magdalene, Bridgnorth.

areas to the north-west of the building site. Eventually, a suitable stone was discovered amongst the fine-grained, current-bedded and pale-coloured Keele Beds outcropping southwards in the township of Eardington where the Mor Brook drains across Carboniferous sandstones.

This stone is ideal for the classical style of architecture that Telford chose for his church. It can be smoothly cut and dressed to ashlar refinement. Much of the external walling masonry is of precisely cut and squared blocks set with fine joints. According to the building specifications of 17 October 1792, each block had to be at least six inches in thickness and all the bond stones had to be set apart at distances of no more than six feet. Most of the walling was given a smooth finish so that any textural variation depended entirely on the current bedding within the natural stone. On the west front and on the lower stage of the tower, a smooth type of rustication was introduced. This involved the sinking of the margins around the individual blocks of stone so that when they were juxtaposed the joints were emphasized by shadowing which produced a boldly textured effect (Fig. 7).

Elsewhere on the church, precision masonry takes the form of smoothly turned Tuscan columns (west front and tower) and Doric pilasters and round arched windows (north and south walls). All these elements are clearly defined despite a lapse of almost two hundred years and they testify to Telford's practical knowledge of stone and his insistence on high standards of masonry craft. The apse at the east end was the work of Arthur Blomfield in the 1870s. He considered the inside of Telford's church to be 'peculiarly cold and depressing' and had the chancel enlivened with stained glass by Messrs Wailes of Newcastle-upon-Tyne in a typical nineteenth-century reaction to a coolly classical church.

An inspection of the present condition of the two Bridgnorth churches of St. Mary Magdalene and St. Leonard's is adequate confirmation of the superiority of the Carboniferous sandstone. The Keele Bed Sandstone also features prominently in areas to the north, especially in and around the town of Shrewsbury. Structures incorporating this type of stone which, like other sandstones noted, has red and paler-coloured versions, include the Roman bathhouse at Wroxeter, Wroxeter church, Shrewsbury Castle and St. Mary's, Shrewsbury. Some of the stone for Condover Hall was quarried at Burriewood on Keele Bed Sandstone although the window mouldings and quoins were executed in Grinshill stone. The medieval town walls of Shrewsbury incorporate Keele Bed Sandstone for this is the rock formation on which the town is situated. It was excavated at the Quarry near to the present site of St. Chad's church and also at Belvidere and Emstry on the outskirts of the town. Much of the sandstone which features on town buildings is a deep red version of the stone.

To the south-west of Shrewsbury, quarries in the Hanwood, Red Hill and Hookagate locality produced a very deep red Carboniferous sandstone that was much in demand during the nineteenth century when Victorian architects were attracted to brightly coloured materials. Holy Trinity, Meole Brace, built 1867-8 to a design of the Shrewsbury architect, E Haycock junior, had rock-faced Red Hill stone for the main walling masonry although, as previously noted, Shelvock freestone was used for contrasting dressings. The engineers of the Shrewsbury and Welshpool Railway which opened in 1862 made use of big blocks of purple-tinted red sandstone obtained from quarries at Stoney Stretton to the west of Hanwood for the construction of a number of railway bridges in the vicinity. Carboniferous sandstones invariably made good engineering stones.

In the Shropshire coalfield east of Shrewsbury, Carboniferous sandstones were useful for some building and engineering purposes although bricks and, to a lesser extent, iron were more important as building materials. The Thick Rock of the Upper Coal Measures was quarried at Malins Lee and Hinksay to the east of Dawley for local building use and the Big Flint Rock of the Middle Coal Measures for bridge construction on the Wellington and Coalbrookdale Railway. Holy Trinity, Dawley Magna, is located on the

Coalport Beds of the Upper Coal Measures, three-and-a-half kilometres north-east of Ironbridge where beds of thick sandstone are interbedded with marls. The Perpendicular-style church which was designed by Harvey Egington of Worcester in 1844 was built of the local sandstone in hammer-dressed blocks laid to regular courses. These Coalport Beds are the geological equivalent of the Highley Beds of the Bridgnorth district.

The remaining Carboniferous areas include that of the Clee Hills in the south of the county where the sandstones of the Carboniferous System are more limited in extent and importance than those of the Old Red Sandstone System. The Cornbrook Sandstone, for example, only outcrops in the Farlow-Oreton-Prescott area to the north-east of Titterstone Clee Hill. This local version of Millstone Grit is inferior to Sweeney Mountain stone, and is far too coarse, friable and pebbly to be of any commercial value. The Coal Measure sandstones of the Clee Hills, although finer-grained, were also restricted to local use on rural buildings.

This brief review of the Carboniferous sandstones of Shropshire has highlighted the most important whilst virtually ignoring a few such as the Coed-yr-Allt Sandstone which was of some significance in the Shrewsbury and Oswestry districts. Except for the Highley and Alveley quarries and, to a lesser extent, the Sweeney Mountain quarries, there was not so much commercial development of the Carboniferous sandstones as in the Denbighshire area of Wales to the north of Oswestry. Most of Shropshire's Carboniferous sandstones were used in their immediate districts of outcrop. This local emphasis added to the diversity of stone-built premises within the county and for this reason alone is important.

7
Sandstones of the Ancient Seas

Sandstones of the Precambrian, Cambrian, Ordovician and Silurian Systems are the oldest in the world, pre-dating the Old Red Sandstone rocks described in Chapter 5. The Precambrian Period reached back in time to the origin of the earth which is believed to be about 4,600 million years old. Britain has its oldest Precambrian rocks in Scotland; those in Shropshire, though relatively young, are still of great antiquity.

The Longmynd Plateau constitutes the most extensive area of Precambrian sedimentary rocks in the county. These rocks include sandstones, grits and conglomerates which can be found on widely dispersed rubblestone premises. In this hilly and remote part of the county, communications were so difficult in former times that local building materials had to suffice. Because of a shortage of timber and clays the region depended on stone. As this was usually difficult to cut and dress it was worked into rubblestone walling. Thick courses of mortar compensated for the assorted shapes and sizes of the constituent stones. Sometimes the rather uncouth masonry was hidden beneath a rendered and limewashed surface which had the additional merit of weatherproofing the walls. Limewashing was also a means of disguising some dark and sombre-looking stones. It gave an illusion of uniformity to walling made up of a motley collection of different stones. The tiny church of St. John Baptist at Myndtown provides an excellent example of a remote, roughcast and limewashed church.

As many of the Precambrian sandstones and grits were tough, unyielding rocks, problems arose with the finish to doors and windows which could not be executed in rubble. If possible, and obviously this was more likely at a later date, the stones used for neatening off openings in buildings were brought in from other regions. When St. Michael, Smethcott, was largely rebuilt in the mid-nineteenth century, greenish-grey grits and sandstones from a local quarry at the northern end of the Longmynd plateau were found to be adequate for the general walling masonry but, as previously noted, Hoar Edge Grit had to be introduced for dressings and Grinshill freestone for window tracery. Few of the churches and chapels in the remotest parts of the Longmynd region have any tracery at all. Most were designed with simple lancet windows and rarely had undercut details on door and window openings. Local rocks were useless for intricate work.

It is believed that the Longmyndian sediments were deposited in the shallow waters of a subsiding trough so that very thick deposits of sandstones, grits, conglomerates and shales accumulated. Subsequent major disturbances of the earth's crust caused these thick sediments to be folded, compacted and generally toughened to such an extent that they could never be exploited by man, many millions of years later, as horizontally disposed and therefore easily quarried layers of freestone.

The catastrophic crustal movements took place prior to the Cambrian Period which commenced about 570 million years ago and was characterized by the invasion of a sea which caused renewed deposition of marine sediments. In Shropshire these were mostly of shallow water type consisting of sandstones, grits and conglomerates, but in Wales vast quantities of finer sediments such as shales and mudstones accumulated in the deeper parts of the sea. These were later converted into slates. The Cambrian sandstones of Shropshire were quarried locally for building stone for vernacular premises, for example, in the Comley district to the east of Church Stretton.

During the Ordovician Period, which is dated between 510-445 million years ago, marine sedimentation prevailed once more with shallow water deposits occurring to the east of what is now the Church Stretton district.

The remote church of St. John Baptist, Myndtown.

Progressively deeper water sediments were deposited in the direction of Wales. Instability in the earth's crust led to much volcanic activity which left a legacy of tough volcanic rocks and some useful mineral deposits of lead, zinc, copper and barytes — notably in the Shelve and Snailbeach localities. These lie to the west of the Stiperstones, a ridge formed out of quartz-rich sandstones which have been toughened into an extremely hard quartzite. The proximity of this shallow water sediment to deeper water sediments such as shales and siltstones can be explained by the ever-changing conditions of the sea bed which resulted from further movements of the earth's crust during late Ordovician times.

The quartzite was obviously a difficult rock to work but it was used as building stone in the remote uplands to the west of the Longmynd. At Ratlinghope, for example, it appears in the walling masonry of the church of St. Margaret. In the north wall the quartzite is dispersed amongst a variety of sombre Longmyndian stones in a completely random arrangement. The south wall which was probably rebuilt in the seventeenth century — the porch door is dated 1623 — is constructed of Stiperstones Quartzite but here, difficult thought it must have been at the time, the tough rock was broken down into squared blocks and laid to level courses with the larger stones at the base of the wall and smaller stones beneath the eaves.

The Ordovician sandstones outcropping to the east were very much easier to work. The Hoar Edge Grit which forms the basal bed of the Ordovician System produced some excellent building stone from amongst its massive, gritty sandstones which are buff-brown or pale grey-white in colour. The flaggy beds produced the roofing 'slates' for some famous Shropshire buildings, including the Prior's Lodge at Much Wenlock and Stokesay Castle. There are several quarries on Hoar Edge itself and also in the district between Cwms Farm and Hoar Edge. Other quarries are found on both sides of the Onny valley to the south-west. Most of these quarries are small and were only exploited for local use in the construction of farm buildings, cottages and churches. Some of the late eighteenth-century repairs to St. John Baptist, Stapleton, involved the use of Hoar Edge stone and for the building of St. Andrew, Hope Bowdler, in 1862-3, S Poutney Smith specified the use of selected Hoar Edge stone for the steps to the entrance, chancel and altar. The stone must have had a sound reputation in the county during the nineteenth century for it appeared with leading Shropshire stones in lists of the nation's main building stones and it featured amongst the British and foreign building stones exhibited at the Sedgwick Museum in Cambridge.

Several useful building stones occur in the geological formation known as the Chatwall Flags and Sandstone. They are named after their respective districts of outcrop. The Horderley Sandstone is found in the south between Sibdon Carwood and Whittingslow. It is a massive sandstone which is coloured green and brown and sometimes banded with purple streaks. South of the Onny valley, quarries in the Longville plantation and in Long Lane were exploited for the purple and brown sandstones which feature on local buildings. Horderley stone is evident in villages such as Wistanstow where the school, built in 1859 to an Italianate design by J Randal of Shrewsbury, has a

Random-coursed walling stones and tooled window voussoirs of local Horderley stone used on the village school at Wistanstow.

combination of roughly squared, random-coursed walling masonry and door and window voussoirs cut from carefully dressed stones with drafted margins.

Hope Bowdler, a village situated some eight kilometres to the north-east of Wistanstow, is built of stone from Soudley Quarry which is located just half a kilometre to the south of the village church. Soudley stone is a massive fine- to medium-grained sandstone which is current-bedded and banded with purple, olive-green and brown colouring. It is a good freestone, much used in the locality and easily identified on the walling of the church.

Further north and quarried in the vicinity of Chatwall Hall, the sandstone — known here as Chatwall Sandstone — was used on rural buildings in the immediate neighbourhood. Other sandstones which are often interbedded with siltstones, shales and limestones of the Ordovician sequence were also quarried for local use on a limited scale.

The Silurian Period which followed between 445-395 million years ago was also a major phase of marine deposition with, once again, accumulations of deep water sediments towards the west and shallow water deposits in the east. Shelf seas covered the Shropshire area in the Much Wenlock, Corvedale and Ludlow districts leaving behind a variety of rocks including conglomerates, shales, limestones and sandstones. The limestones provide useful clues concerning the geographical position of the British Isles in Silurian times. The reef corals trapped within the Wenlock Limestone, for example, suggest a warm and clear shallow water environment within tropical latitudes somewhere just south of the equator. The earth movements that had taken place at

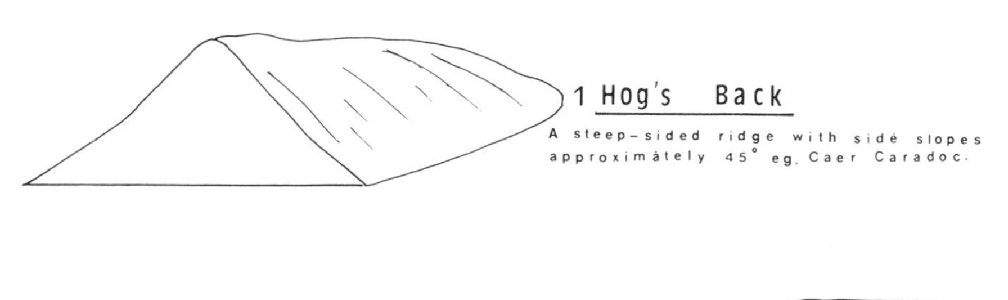
1 Hog's Back
A steep-sided ridge with side slopes approximately 45° eg. Caer Caradoc.

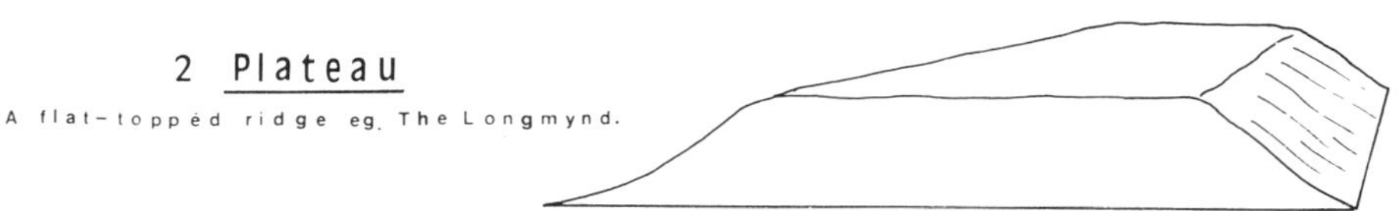
2 Plateau
A flat-topped ridge eg. The Longmynd.

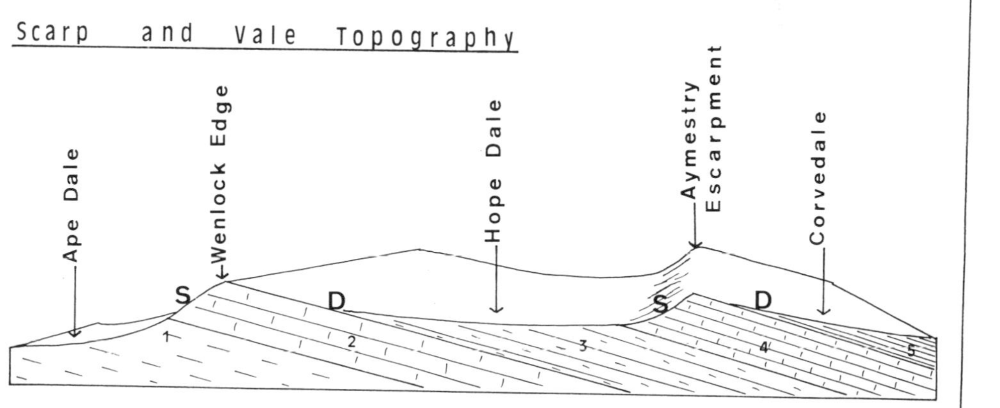
Scarp and Vale Topography
Ape Dale
Wenlock Edge
Hope Dale
Aymestry Escarpment
Corvedale
S
D
S
D
1
2
3
4
5

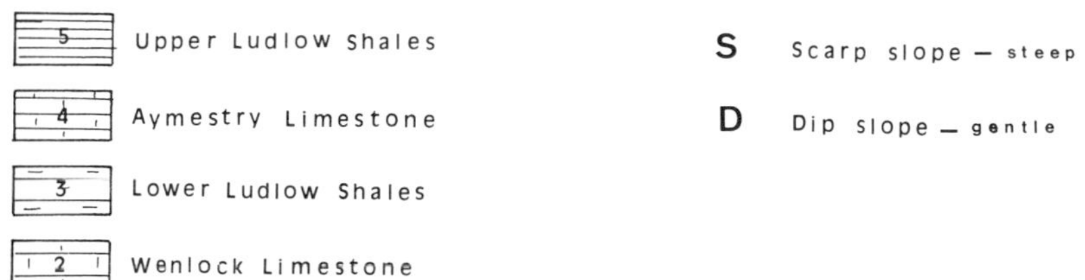
OCK SEQUENCE:-
5 Upper Ludlow Shales
4 Aymestry Limestone
3 Lower Ludlow Shales
2 Wenlock Limestone
1 Wenlock Shales
S Scarp slope — steep
D Dip slope — gentle

Diagrams not to scale.

Figure 8: Landforms

the end of the Ordovician Period had caused the sea to retreat westwards but it returned to the western side of Shropshire during Silurian times when marine life was especially abundant. The Silurian rocks are therefore highly fossiliferous. Some of the sandstones and grits were important locally as building stones, for example, Kenley Grit features in the churches at both Kenley and Harley. It was, however, the limestones that had most economic significance, though less so for building stone purposes than as a source of agricultural and industrial lime. However, mortar was an important by-product of limestone quarrying of considerable use to the building trades.

Major earth movements at the end of the Silurian Period had widespread effects in Britain. Folding and faulting movements had most impact in Scotland and Wales but the Shropshire area was also affected. Synclines (downfolds) formed in the Clun Forest and Long Mountain areas; anticlines (upfolds) in the Ludlow district (Fig. 6). East of Church Stretton, a gentle tilting of the rocks towards the south-east set the stage for the gradual evolution of the scarp and vale topography that is so striking today in the Wenlock Edge and Corvedale region (Fig. 8).

The dominance of marine conditions during the Precambrian, Cambrian, Ordovician and Silurian Periods ended with the encroachment of mainly continental conditions during the Old Red Sandstone Period. These in turn gave way to a major incursion of the sea during Carboniferous times but in the succeeding Permian and Triassic Periods continental conditions returned with a vengeance and resulted in the formation of widespread deserts. The sandstone legacy to be found in Shropshire at the present day shows concentrations of continental desert sandstones in the north and east of the county and marine sandstones in the south and west, but this is a very generalized picture.

In considering Shropshire's building stone endowment, full credit must be given to the sandstones and related grits since most of the county's major freestone quarries were located on these outcrops. A few select sandstones supplied those freestones which were marketed successfully beyond the county boundaries. In addition, they featured on a wide range of buildings within the county, some of which are now regarded as architectural showpieces. The little sandstone quarries which were opened up for local use have also made a significant contribution to Shropshire's diversity of stone buildings. The other major types of rocks found in Shropshire never supplied such quantities of building stone but they merit attention because of their impact on local vernacular architecture. The remaining chapters deal with rocks other than sandstones.

8
Limestones for Building Use

Limestones have never been as important as sandstones for building stone purposes in Shropshire. This is partly explained by their more restricted geographical distribution but, more significantly, they were quarried primarily as a source of agricultural and industrial lime. Nevertheless, a few districts are characterized by limestone buildings and lime mortar was widely used in the county in previous centuries. Shropshire limestones are now mostly utilized as roadstone.

The particular types of limestone to be found in Shropshire have little in common with the famous oolitic limestones of the Cotswolds which yield freestones of Jurassic age. Shropshire's limestones were mostly laid down in the Silurian and Carboniferous periods, with the older of these two featuring thinly bedded, nodular limestones and the younger the massive, tough, crystalline limestones which, being closely identified with rugged uplands such as the Pennines, are often referred to as 'mountain limestones'. Shropshire's limestones are mostly found on rubblestone premises because they are generally unsuitable for ashlar masonry.

It is useful to consider some general features of limestones in relation to building purposes before specific stones and related buildings are described. An essential ingredient of limestones is calcium carbonate which varies considerably in proportion to other minerals. The purest limestones contain eighty to ninety-eight per cent carbonate of lime and impure limestones less than thirty per cent. The lime is derived from diverse sources including shell fragments and skeletons as well as chemical precipitations. Limestones therefore vary widely in terms of chemical content, texture, hardness and colour. Some are highly fossiliferous and anything but smoothly textured. Tufa, as previously described, has a rough, cellular texture but it is surprisingly tough. On the other hand, fine-grained, even-textured limestones such as chalk may be relatively soft. Some limestones are very clayey (argillaceous) whilst others are so sandy (arenaceous) that they are difficult to distinguish from calcareous sandstones. Carboniferous limestones are mostly dense, hard and crystalline — so tough in fact that they are difficult to cut and dress. A few are sufficiently hard and compact to be polished and used ornamentally as a substitute for marble, which is really a metamorphosed limestone that has been so much altered by the heat and pressure associated with major earth movements that any original fossils it may have contained are entirely obscured. In the polished sedimentary limestones, the fossils remain conspicuous and create some interesting patterning, much exploited by the manufacturers of fireplaces and ornamental cladding. Many of the Silurian limestones are nodular, that is they contain assorted lumps of harder material.

The durability of limestones as building material is at least as variable as that of the sandstones already described. The mineral content and the state of aggregration of the constituent minerals are key factors which interact with environmental factors in such complex ways that it is virtually impossible to generalize. For example, a limestone that survives well in country districts may decay relatively rapidly in smoky urban/industrial locations where rainwater carries sulphuric and hydrochloric acids, in addition to the carbonic acid normally washed out of the atmosphere. When calcium carbonate recrystallizes into calcium sulphate the increase of crystal volume results in the blistering and eventual peeling away of the faces of building stones. The porous freestones of the Bath area are very prone to this type of attack; Shropshire's limestones are generally less so since they are usually harder and more crystalline. However, it is possible for a sharp frost to split individual calcite crystals so that an inflow of acid-laden water can cause the decay of even the hardest limestones. Sometimes the softer, more porous limestones are actually protected from weathering because their more open texture permits a relatively rapid release of moisture. Lingering damp is the most destructive.

The decay of limestone masonry is closely affected by its position in a building. Sheltered stones lying underneath projections may harbour damp and are therefore very vulnerable to chemical weathering and some frost damage. Exposed portions of a structure such as cornices and string courses are most prone to physical attack such as wind attrition, temperature changes and frost. All stones weather with time but at very different rates. Some stones wear back evenly but others so unevenly that blotchy patches form on the masonry. Limestones with a high proportion of clay wear away very patchily and not at all decorously.

In Chapter 7 reference was made to the presence of shallow-water shelf seas over western Shropshire during the Silurian Period. These were the origin of some highly fossiliferous limestones, notably the Wenlock and Aymestry Limestones. Present-day outcrops are concentrated south of the Severn especially in the Ludlow, Craven Arms and Much Wenlock districts but they also reach across the river to Lincoln Hill which was exploited for the flux used in the iron-making industry of the East Shropshire coalfield in the eighteenth and nineteenth centuries.

As a building stone, Wenlock limestone had limited importance in terms of its distribution but as a source of mortar it was used in many distant parts of

Shropshire. The reputation of Wenlock lime mortar was upheld not only by county architects and builders but also by national architects issuing their building instructions from London offices. R H Carpenter and W B Ingelow specified 'best Wenlock stone lime' for the mortar used in the painstaking restoration of the tower of St. Mary's, Market Drayton, in 1885.

Wenlock Limestone has all the essential characteristics of a rock formed in warm, tropical seas, which in this case steadily encroached from the west in Silurian times. Coral reefs were able to grow in the clear stretches of water. On Wenlock Edge in the vicinity of Presthope, the limestone is crammed with banks of corals, bryozoa, crinoids and stromatoporoids which flourished in the warm, sunlit waters. Elsewhere, muddier waters left a legacy of rubbly limestones which contain a scattering of solitary corals. Much of the Wenlock Limestone is thinly bedded and nodular and therefore totally unsuitable for smoothly dressed masonry. Its use was confined to local vernacular rubblestone buildings constructed with considerable expertise by local masons who were well practised in the art of assembling walls from awkwardly shaped stones. Large quantities of mortar were needed but this was obviously no problem in a limestone district. The limestone quarries providing stone, lime and also flux for the Lilleshall Company were concentrated in the Much Wenlock area and in the hills stretching north-eastwards towards Benthall Edge. An elaborate transport system which included inclines, tramways, water and later rail haulage evolved from the late eighteenth century in response to the needs of the iron-making industry located on the coalfield to the north.

The Aymestry Limestone outcrop parallels Wenlock Edge on the east, providing villages such as Aston Munslow, Munslow and Diddlebury with rough rubblestone for cottages and farmhouses. It also outcrops in the Onibury district, both west of the Onny at View Edge, where the brachiopod Conchidium Knightii is conspicuous in the quarry faces, and east of the Onny at Whettleton Bank and Norton Camp. The nodular, crystalline limestone is often interspersed with siltstone partings, denoting unstable conditions of sedimentation. Most quarries could only supply thinly bedded, rough limestone of limited local significance.

Shropshire's Carboniferous limestones are found both north and south of the Severn having been deposited in two distinct basins which were probably separated by dry land. In the northern basin the limestones, which include those of the Llanymynech-Oswestry district in the west and the Lilleshall district in the east, were deposited as thick accumulations, but south of the Severn only restricted outcrops are found, mostly on the north-eastern flanks of Titterstone Clee Hill. Here, amongst the ordinary limestone which was used as rough building stone and as lime for mortar and agricultural use, a distinctive type of grey-blue, shelly oolitic limestone was extracted for polishing and marketing as 'Clee Hill Marble'. As it was full of crinoid fossils it had an interesting patterning which recommended its use as an ornamental stone, especially for the decorative fireplaces and the classical columns featured in local country mansions.

The economic motivation for quarrying the Carboniferous Limestone outcrops north of the Severn came from the iron industry and from agriculture. Pastoral farmers in Central Wales and Cheshire needed lime to improve their land, so lime was burnt in numerous kilns along canal routes leading to these two regions. Once the iron manufacturers of East Shropshire had exhausted the limestone supplies of the Lilleshall district, they turned their attention to the hills around Llanymynech on the Welsh border. Here the limestone outcrops represent the southern part of the elongated sequence of Carboniferous Limestone that forms some of the most dramatic scenery in Clwyd, notably the impressive Eglwyseg Escarpment north of Llangollen. The Ellesmere Canal with its Llanymynech Branch, which was linked to the Montgomeryshire Canal at Carreghofa in the late eighteenth century, was the main transport artery until the Cambrian Railway and the Potteries, Shrewsbury, and North Wales Railway were built in the following century. Self-acting inclines, tramways and light railways served the Llanymynech district as a means of conveying the limestone from the hillside quarries to the mainline communications.

The western outcrop of Carboniferous Limestone was also important as a source of building stone since, unlike the limestone outcrops previously described, it contained massive beds of stone with the potential for supplying the kind of large dimension blocks needed by the engineer and builder. This was especially so in the case of the Upper Grey Limestone beds which comprised a very hard, crystalline limestone which was difficult but not impossible to cut and dress. Many of its squared blocks were incorporated into the fabric of local churches and larger, more imposing farmhouses. Productive building stone quarries were located in Treflach Wood where corner stones, gateposts and kerbstones were also cut out of the thick beds of the Upper Grey Limestone. Underneath the best building stone beds lay very fissile layers of limestone which were ideal for splitting into gravestones and paving slabs, whilst above, the distinctive ‘Coral Bed’ was used as an ornamental stone. Like the ‘Clee Hill Marble’ previously referred to it was polished and marketed for monuments and fireplaces. It was known as ‘Welsh Marble’.

Numerous limestone quarries scattered across the hills to the west of Oswestry supplied rough rubblestone for the construction of rural premises and chapels in the nineteenth century. Both the church and the parsonage at Rhyd-y-Croesau incorporated walling stones of rubble obtained from nearby quarries but the dressing stones for the parsonage came from Treflach Wood limestone quarries and for the church from sandstone quarries yielding a more easily cut buff-coloured stone. Most rural bridges were also built of rubble limestone but on rare occasions, when an architect designed a more refined and stylish bridge, the limestone was laboriously squared and dressed. Pontfadoc Bridge, designed by Edward Haycock in 1836, spans the Morda to the east of Llanymynech and though in an open, rural location it was deemed worthy of precision masonry work by a local Oswestry mason, Richard Milnes. Precisely cut chamfered voussoirs were made from the tough Llanymynech limestone.

On the eastern side of Shropshire, the outcrops of Carboniferous

Limestone were used relatively less as building stone because of the urgent demands of industry for limestone flux. Bricks made from Carboniferous clays in addition to the local Carboniferous sandstones were more important building materials in the industrial districts of the coalfield. However, limestone was the main walling material used in the construction of St. George's, Oakengates, which was designed by the architect G E Street in 1861. As it was to be the memorial church to the late Duke of Sutherland it was convenient to use stone from the limestone quarries on the family's Lilleshall estate. This was adequate for the rough-textured and random-coursed walling masonry but the dressings were cut from quality freestone obtained at Grinshill.

Limestones other than those associated with Silurian and Carboniferous sedimentation were used as building material in various parts of Shropshire. The Ordovician System includes the Acton Scott Limestone — a hard, sandy (arenaceous) type of limestone which outcrops on the high ground around the village of Acton Scott in South Shropshire. Several quarries close to the hall and the church were exploited for the limestone which has weathered into the grey-brown colour prevalent on stone-built premises in the village.

The role of limestones in the architectural heritage of Shropshire is primarily associated with vernacular premises in rural locations. Here, rubblestone construction suited the functions of the buildings and the limitations of the stone. The various limestones could never compete with sandstones as building stone essential for the construction of more complex and refined styles of architecture so their range was virtually limited to modest structures in rural areas. For this reason they add character to the local scene and contribute to the diversity of the county's architecture.

9
Miscellaneous Building Stones

Sandstones and limestones of different geological ages constitute the bulk of the building stone used in Shropshire but other sedimentary rocks, namely mudstones, shales, siltstones and breccias have been used in certain localities. Unlike some of the stones described in previous chapters none of these had any commercial value beyond their respective districts of outcrop.

The individual particles in mudstones, shales and siltstones are often so minute that they can only be identified through a powerful microscope. These fine-grained rocks are clayey (argillaceous) and soft and smooth to the touch although the siltstones, with their relatively high proportion of tiny quartz grains, may feel slightly rougher.

Clay particles may be suspended in water for a long time and they can only very gradually settle down in still or gently moving water. Most deposition occurs in lakes and the deeper parts of oceans so that Wales, which was part of a deep ocean basin during Cambrian, Ordovician and Silurian times, has massive accumulations of such fine-grained deposits. Western Shropshire shared these ancient environmental conditions so it is in the Clun Forest region, the upland district to the west of the Longmynd and, to a lesser extent, the Scarp and Vale region east of Church Stretton where the seas were somewhat shallower, that most of these fine-grained rocks are found. When clay is compacted into mudstone much of its water content is eliminated so that a dense mass of rock with few bedding planes is formed. Shales are similar in composition but because their bedding planes are closely spaced they can easily split into thin layers. Siltstones are slightly coarser in texture but they resemble mudstones when thickly bedded and shales when thinly bedded. All three types of rock constitute indifferent building stones but despite this they feature on a surprising number of buildings, some of which involved the work of architects of national repute.

During the Silurian Period the deposition of the limestones referred to in the previous chapter alternated with the deposition of the fine-grained rocks just described. The differential erosion of the more resistant limestones and the softer shales has resulted in the Scarp and Vale landscape shown in Fig. 4. The relative resistance of the rocks is also manifest in the way they have succumbed to weathering as building stones, with the limestones proving to be far more durable. The present day use of Wenlock Limestone as roadstone confirms this toughness. The shales, mudstones and siltstones were obviously much easier for the masons to work and therefore tempting to use as building stones but some of them have weathered badly, causing severe problems with the masonry fabric of many buildings in Corvedale and in the Ludlow and Clun Forest districts. The siltstones used on St. Laurence, Ludlow, have already been cited in Chapter 5. Other buildings in the vicinity of Ludlow Castle also show the inadvisability of using soft, thinly laminated siltstone.

Along Corvedale, Upper Ludlow Shales outcrop on the lower dip slope of the Aymestry escarpment and in the vale to the east. Several churches in this locality incorporate walling stones cut from the olive-green siltstones and shales. The severe weathering of these stones is apparent on the porch of St. Peter, Diddlebury, where the contrast with the relatively pristine condition of the red sandstone blocks used on the quoins and on the window surrounds accentuates the frailty of the siltstones. An even more pronounced weathering of similar stones at the base of the walls of Holy Trinity, Bourton, is largely due to rainsplash erosion. St. James, Shipton, and St. Michael, Munslow, also have walling stones of badly weathered grey-green siltstones and shales. On the tower of the latter church, an attempt at rendering has highlighted the dilapidated condition of the masonry fabric beneath.

In the Clun Forest region of south-west Shropshire the rocks are derived from the argillaceous and silty sediments characteristic of the deep-water

Severe weathering of siltstones at the base of a wall at Holy Trinity, Bourton.

deposits of Wales and are less calcareous than their counterparts in the districts just described. Mudstones and siltstones prevail. Despite their dubious merit as building stones they feature on many premises including churches because there were so few alternative materials available in this remote part of the county. Two outlying churches built of local siltstones in the mid-nineteenth century, St. John the Evangelist, Newcastle, and St. Mary, Llanvair Waterdine, provide ample evidence of the poor weathering potential of the stone. Similar decay is visible on the south aisle wall of St. George, Clun, which was built during G E Street's restoration of this ancient church in the 1870s.

The new church of All Saints, Richard's Castle, which was constructed 1891-2 to a design of R Norman Shaw, the architect of Adcote, has external walling masonry of local Silurian siltstones which were quarried on the Moor Park estate of Shaw's wealthy client, Mrs Hannah Foster. Grinshill stone was used for the dressings and also for the internal ashlared walling. The contrast between the two types of stone is all too apparent where they are juxtaposed around the entrance porch and windows and at the corners. The siltstones, being finely laminated and relatively soft, have split and peeled in so many places that flakes of stone now litter the base of the church.

In complete contrast to the fine-grained rocks just described, the Alberbury Breccia, which outcrops mostly on the Loton estate some twelve kilometres west of Shrewsbury, is a rough-textured, coarse-grained, lumpy type of rock known locally as Cardeston Stone. It was used as building material on a number of farms and cottages in the district, proving especially

These finely-laminated siltstones in the walling of All Saints, Richard's Castle, split and flake fairly rapidly.

useful in the nineteenth century when improvements were being made to premises on the Loton estate. Breccia was also used in church construction and restoration.

Breccia is a sedimentary rock which evolved in conditions totally at variance with the quiet, steady sedimentation necessary for the formation of fine-grained and evenly-textured rocks. In this respect it is similar to conglomerate but it contains a larger proportion of angular fragments. The Alberbury Breccia is made up of quartz pebbles, red marl and angular fragments of Carboniferous Limestone which are believed to have come from outcrops to the south of what is now the Llanymynech district. During Permian times, when the climate was becoming progressively drier, these limestone uplands provided some of the detritus washed into the developing desert basins by the flash floods resulting from sudden, torrential desert rainstorms. As these subsided, masses of rough scree were dumped around the edges of the basins along with sands and clays. This unsorted debris was destined to become the Alberbury Breccia — an exceedingly tough rock comprising pebbles and lumps of broken rock fragments welded into a calcareous, sandy matrix.

Obviously no mason could attempt any ashlaring of such an unyielding stone so it was used in rubblestone walling with ample applications of mortar. Cottages and farm buildings and the occasional school, such as Wattlesborough School, invariably had dressings of brickwork around the doors and windows but churches, which were usually constructed of roughly squared breccia stones laid to random courses, had dressings of quality freestone. The church of Holy Trinity at Bicton has already been mentioned in Chapter 3 in connection with the use of Shelvock stone dressings. On this building the breccia, which constituted the bulk of the walling masonry, was squared but left rough on the face in contrast to the smoothly dressed freestones.

The rural district to the west of Shrewsbury contains several churches incorporating masonry of Cardeston stone. Examples include Holy Trinity, Yockleton, which was built in 1861 to a design of E Haycock, who used the rough-textured breccia stone for the walling and Grinshill stone for the dressings. Both the 'red' and 'white' varieties of Grinshill were used for decorative impact on the belfry and the dormer windows of the tower. Other churches where breccia can be identified on the external fabric include St. Michael, Cardeston, St. Michael, Alberbury, St. Michael, Ford, and St. John, Wollaston.

The use of such contrasting rocks as shale and breccia reflects the dependency of many local communities on the geological bounty of their immediate surroundings. The expendiency and economy of quarrying local outcrops often overruled any doubts about the durability of the stone and, in any case, deficiencies were not always apparent at the time. Unlike the other stones previously described in this chapter, the breccia has proved to be a sturdy stone and one well suited to a rural environment because of its inherent roughness and rusticity.

Because the outcrop of Alberbury Breccia is so limited in extent it lends character to relatively few stone-built premises in a small area of Shropshire,

The battlemented tower of Holy Trinity, Bicton, is sturdily built of rough Alberbury Breccia dressed with sandstone.

but in so doing it contributes a disproportionate amount to the diversity of architecture within the county. The same may be said of the igneous and metamorphic rocks outcropping in Shropshire, the formation of which depended on a complexity of factors involving fundamental disturbances within the body of the earth.

Briefly, igneous rocks evolve from the cooling of molten material which exists in the mantle beneath the solid crust of the earth. Movements within the molten mantle cause ruptures and collisions in the crust which 'floats' on top. On rare occasions the molten material bursts out in volcanoes and lava flows to form new rock on the surface. Very rapid cooling results in amorphous glass-like rocks such as obsidian but most igneous rocks formed at the surface have tiny crystals. Some of the molten substance, however, cools slowly at depth so that large crystals have time to develop — the deeper the location of cooling the larger the crystals. Granite is believed to form at depth and only outcrops at the surface of the earth after millions of years of denudation of the covering rocks. Most igneous rocks are tough and very difficult to work. In the nineteenth century the masons and quarrymen working granite were paid higher wages than their counterparts in the sedimentary freestone quarries and granite has always been one of the more expensive stones.

Shropshire is not a county renowned for granite although the term has been applied to the Clee Hill dhustone which is really a fine-grained and dark-coloured olivine-dolerite. A superb roadstone with a sound nationwide reputation at the turn of the century it was never a building stone of any significance except in proximity to the coalmines and quarries of Titterstone Clee Hill where the dhustone was conveniently located for some opportune construction of squatter cottages. Similar crudely and quickly erected dwellings incorporating dolerite boulders can be found in the Minsterley district of western Shropshire.

Another misnomer for Clee Hill dhustone is 'basalt' which is really a fine-grained igneous rock extruded at or near the surface of the crust. In terms of texture, dolerites are intermediate between the fine-grained basalts and the coarse-grained granites, so it is easy to appreciate why both terms are mistakenly applied to a rock formed at depths ranging between the two extremes. Clee Hill dhustone was marketed either as 'granite' or as 'basalt' in

Window detail from Holy Trinity, Bicton, showing how the masons juxtaposed the dressed freestones and the random-coursed breccia walling stones.

Whittery Bridge is constructed of a tough volcanic rock from Whittery Quarry but the arch is neatly turned in four courses of bricks.

the late nineteenth century. The more correctly named Little Wenlock Basalt which is found in the vicinity of the East Shropshire Coalfield close to the Wrekin was also, on occasions, used as rough building stone for cottage construction though roadstone was its main product. In fact all dolerites and basalts are far more useful as roadstone than as building stone, but wherever they outcrop there is usually a scattering of rubblestone cottages which are often limewashed to counteract the rather sombre and forbidding appearance of the stone.

Clee Hill dhustone and Little Wenlock Basalt are relatively young igneous rocks which were forced into and through Carboniferous rocks. A much older igneous rock, found amongst the Ordovician outcrops in the west of the county, owes its origin to a more explosive type of igneous activity. Whittery rock derives from volcanoes which erupted in the Shelve district at a time when Wales and the Lake District were experiencing much more widespread and devastating vulcanicity. It outcrops in Whittery Wood to the east of Chirbury and was quarried for rubblestone walling on local properties and for the rubblestone construction of Whittery Bridge. Because of the extreme toughness of the stone, cutting to any degree of refinement was impossible so that all masonry edges had to be neatened off with other materials. Local domestic premises usually feature brick quoins, door and window surrounds; Whittery Bridge has a precisely finished four-ring brick arch. However, since brick dressings are not so compatible with ancient ecclesiastical architecture, Edward Haycock introduced Shelvock freestone as a foil to the rougher Whittery stone in his restoration of St Michael, Chirbury, in the second half of the nineteenth century. The masonry techniques adapted to the inherent characteristics of these two contrasting rocks are shown in the photograph of St. Michael, Chirbury (see first picture of colour section).

Metamorphic rocks are formed from existing rocks to which either heat or pressure and sometimes a combination of both are applied so that sedimentary, igneous and even pre-existing metamorphic rocks are altered to some degree. The Stiperstones Quartzite, described in Chapter 7 along with other Ordovician sandstones, is in reality a metamorphosed sandstone. Low-grade metamorphism leaves the rocks with many of their original characteristics; high-grade metamorphism completely alters the rocks, and generally toughens them so much that they are most difficult to quarry and work. Sandstones tend to have their quartz grains bonded into such a tight interlocking mass that the resulting quartzite is a brittle stone. In the natural environment, weathering in general and frost damage in particular causes the rock to shatter into the kind of sharp-edged fragments visible on the Stiperstones ridge. In complete contrast, some metamorphosed limestones change to marble — a most consistently fine-grained and easily tooled stone much prized by monumental sculptors down the centuries. With few exceptions, metamorphic rocks are far more useful as roadstone than as building stone.

Quartzite is a metamorphic rock well represented in Shropshire. In addition to the Stiperstones ridge, it can be found in the Nill's Hill, Pontesbury and Granham's Moor districts to the north and also on the flanks of the Wrekin. In all these locations it has been much exploited as roadstone.

A stack of Harnage stone 'slates' ready for roof renovation at Pitchford Hall. Note the fissile nature of the thinly bedded and fine-grained sandstone.

Use of the Stiperstones quartzite as building stone was no doubt encouraged by the lack of competing materials in the district and the fact that this particular quartzite is attractively pale in colour and needs no limewashing to make it presentable.

As Shropshire lacks metamorphic slate it has been customary for centuries to bring in consignments of roofing slates from quarries and mines located just across the Welsh border in the Ceiriog, Tanat and Dee valleys and from further afield in mountainous North Wales. The supply of indigenous sedimentary stone 'slates' could not meet the demand for roofing material in the county although such slates were locally important and featured with distinction on prestige properties such as Pitchford Hall.

In addition, a few igneous rocks were obtained from the border region and

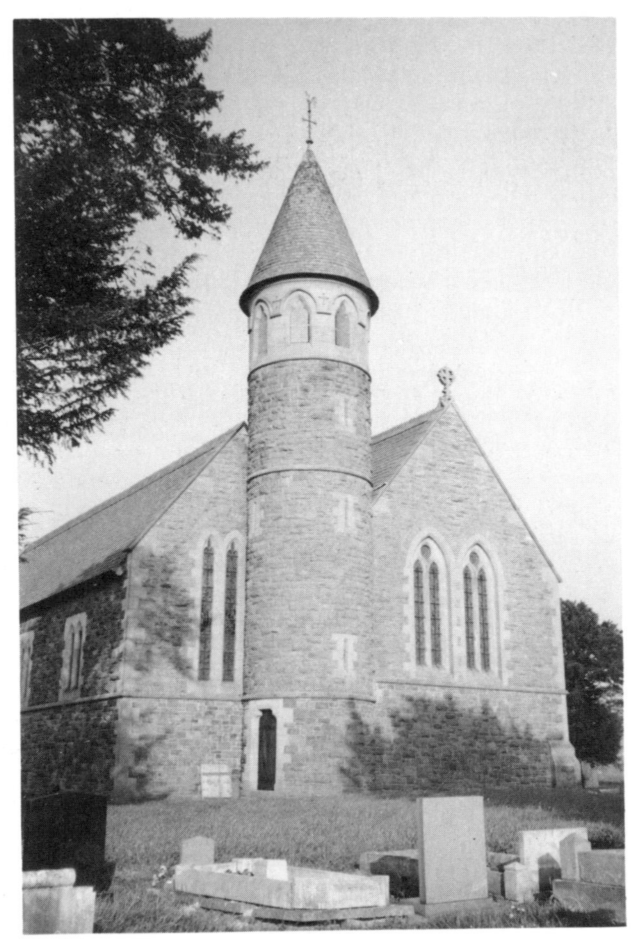

G E Street's St. Tysilio, Llandysilio, with its round tower neatly executed in Criggion Green dolerite and Cefn stone dressings.

Detail of the precision masonry on St. Tysilio, Llandysilio.

featured on premises just within the county boundary. One of the most sought-after was Criggion Green, often listed in the nineteenth century as a stone of visual appeal at a time when a striking colour was considered an asset. Outcropping in the steep-sided Breidden Hills of Montgomeryshire (Powys), it is part of an assemblage of extremely tough rocks that feature dramatically as an upland mass in sharp contrast with the lowland tract to the west of Shrewsbury. The architect, G E Street, better known for his design of the Law Courts in the Strand, London, worked on a number of churches in Shropshire and along the Welsh borders in the last century — a time of great activity in church building and renovation. For the walling of St. Tysilio, Llandysilio, which was built in 1867-8, he used the local Criggion Green dolerite, setting it off against dressings worked in the buff Cefn freestone quarried from Carboniferous outcrops in the Ruabon district to the north of the Dee. The use of the tough Criggion Green stone was bold. Each individual block was accurately squared, though left rough on the face, and was meticulously laid in random courses incorporating rectangular blocks of widely different dimensions. The dolerite can also be seen as rubblestone masonry on a few domestic premises along the border. It is now used as a roadstone, a fate common to most igneous and metamorphic rocks.

The refinements of the masonry work on the church of St. Tysilio are a rare occurrence. Most igneous and metamorphic rocks used in building ended up in crude rubblestone masonry. The preponderance of rubblestone buildings and bridges in the west of Shropshire is explained by the concentrations of tough rocks in the upland regions of the west of the county and in the adjacent

Welsh borderlands. Here, the close link between the natural landscape and man-made constructions is self-evident in the local rocks. It is precisely the toughest rocks that constitute the uplands most resistant to weathering and it is these same rocks that the builder is usually obliged to treat as rubblestone. Though commercially less desirable as building stones, the rocks of this western part of the county are full of character and add to the variety of natural and cultural landscapes which makes Shropshire a county rich in visual interest.

Glossary

ABUTMENT — the solid structure at the sides of a bridge which resists the thrust of the arch.

AEOLIAN — transported by wind.

ALABASTER — a fine-grained mostly white type of gypsum used as ornamental stone.

ANTICLINE — an arched fold in rocks — an upfold.

APSE — semicircular eastern termination of a church.

ARENACEOUS — sandy. Includes sandstones and grits.

ARGILLACEOUS — clayey. Includes clays, shales, mudstones, siltstones and marls.

ASHLAR — neatly cut, squared and dressed stone fixed in regular courses with fine joints.

BACKING — the brick or rough stone walling behind the facework.

BALUSTER — a small pillar, usually with curving, shaped outline.

BARYTES — a barium mineral usually concentrated in veins and nodules in rocks.

BASALT — a fine-grained igneous rock intruded into and flowing over the surface of the earth's crust. It solidifies into a tough rock suitable for roadstone.

BED — a single layer of sedimentary rock separated from others by bedding planes.

BEDDING — a sequence of deposition which may appear as layers of rock separated by bedding planes but can also take the form of gradations of colour and/or texture within a single layer of rock.

BEDDING PLANE — a distinctive parting between layers of rock which results from pauses in the deposition of sediments.

BEDROCK — the solid rock beneath unconsolidated deposits.

BOND — any material which binds separate stones together, e.g. mortar.

BOND STONES — stones fixed at right angles to the outer face of a wall and connecting into the backing to hold the masonry firmly. Sometimes called 'through stones'.

BRACHIOPOD — a bivalved marine animal consisting of two shells.

BRECCIA — a rough type of rock comprising angular fragments of ill-sorted debris.

BRYOZOA — moss-like colonial, mostly marine animals found in limestones.

BURR — rough quarried stone, used mainly for backing and rough walling.

BUTTRESS — a masonry or brick structure built to support a wall against any outward thrust; essential in ecclesiastical architecture to counter the thrust of high vaulting.

CALCAREOUS — containing a high proportion of calcium carbonate.

CAP — the capital or top part of a column or pilaster.

CASTELLATED — with decorative elements such as turrets and battlements.

CHALK — a fine-grained pure limestone.

CHAMFERED — cut off diagonally or bevelled.

CHANNELLING MACHINE — a device for cutting deep grooves into a quarry face to assist in the removal of blocks of stone.

CLADDING — a thin layer of stone, slate, tile etc. covering the surface of a structure.

CLAY GALL — lens-shaped patch of hardened clay within sandstones.

COLUMN — a cylindrical and slightly tapering pillar which may constitute the main part of a monument or act as support to the higher structural elements of a building.

CONCHIDIUM KNIGHTII — a large brachiopod with convex valves which are strongly ridged. Occurs in fossil shell banks on View Edge, near Stokesay.

CONGLOMERATE — a rock made up of rounded and sub-rounded pebbles (which accumulated on former beaches and in river beds) embedded in a sandstone or limestone matrix.

COPING — a projecting capping to a wall — usually with a sloping surface to shed rainwater.

CORALS — mostly marine animals but some are freshwater; some secrete hard skeletons which form reefs out of coral colonies; others are solitary corals.

CORNICE — a projecting element of a structure e.g. above window, door; in classical architecture the cornice is the top member of the entablature supported by the columns.

COURSE — a horizontal layer of bricks or stones.

CRAMP — a metal bar with turned ends for holding masonry. Iron cramps may react adversely with the chemical content of some stones, causing decay. Hence the substitution of slate dovetail cramps.

CRENELLATED — battlemented.

CRINOID — a marine animal; one type resembles a lily in form — often referred to as a 'sea lily' — the fossilized stems are sometimes preserved in limestones which may be polished as ornamental stone.

CROCKET — a carved projection often in the form of a leaf, mostly adorning the angles of spires and pinnacles.

CROSS-STRATIFICATION — a series of sloping bedding planes indicating the mode of deposition of the original sediment. Sometimes termed FALSE-BEDDING.
See CURRENT-BEDDING AND DUNE-BEDDING.

CRUST — the outermost layer of the earth.

CRYSTALLINE — showing the crystal structure of the constituent minerals of a rock e.g. granite made up of quartz, feldspar and mica.

CURRENT-BEDDING — sloping layers of sediment defined by inclined bedding planes which reflect the movement of material by shifting currents.

CUSP — the ornamental projection carved at the intersection of two arcs in Gothic window tracery.

DELTA — an accumulation of sediment at the mouth of a river where it enters a lake or sea with little tidal or current scour.

DHUSTONE — the local name applied to the olivine-dolerite found in the Clee Hills.

DIAGONAL TOOLING — the marks cut on the faces of stones diagonally to the margins. A variant is HERRINGBONE TOOLING.

DIFFERENTIAL WEATHERING — occurs in rocks of unequal resistance so that the more resistant form uplands, the less resistant the vales and lowlands.

DIMENSION BLOCKS — larger blocks of stone of specified sizes, cut at the quarries and in stoneyards usually for monumental architecture.

DOLERITE — a medium-grained igneous rock rich in basic minerals.

DORIC — one of the Greek orders of architecture featuring a sturdy, fluted column. The plain Roman version is usually taller and more slender.

DOWEL — a peg of wood, metal or slate sunk into sockets in two adjacent blocks of stone to hold them together.

DRAFTED MARGIN — the tooled edge on the face of a block of stone, usually about three-quarters of an inch wide. The four margins frame the block.

DRESSING — cutting to a smooth finish. Dressed stone is used to accentuate architectural highlights and for neatening off door and window openings.

DUNE-BEDDING — a distinctive type of cross-stratification with sloping beds following the pattern of accumulation of sand grains in dune formations. Because of the constant changes of wind direction the beds are inclined at different angles in quite complex patterns.

DYKE — an igneous intrusion that cuts through the 'country' or host rock in a discordant way.

ESCARPMENT — a hill ridge with a steep slope and a gently inclined opposing slope.

FACE-SHALING — the peeling of the faces of stones in contact with the atmosphere.

FACEWORK — the outer masonry, usually undertaken by a skilled mason.

FACING — the outer surface of a wall, usually constructed of better quality materials.

FALSE-BEDDING — cross-stratification.

FAULT — a fracture through the rocks of the earth's crust which may cause considerable lateral and/or vertical displacement of the rocks on either side.

FAULT-LINE — the line, sometimes traceable on the ground, where the fault intersects the surface of the earth. Fault lines are drawn on geological maps.

FERRUGINOUS — brown or red coloured, rich in iron.

FINIAL — an ornament at the top of a gable, spire, pinnacle etc.

FIRESTONE — a stone that resists fire and is therefore useful in furnaces, fireplaces etc.

FLAGSTONE — a thinly-bedded rock suitable for flags and paving.

FOLD — a flexure in the stratification of rocks caused by lateral pressure induced by the movement of blocks of the earth's crust. See ANTICLINE and SYNCLINE.

FREE CLASSIC — a style of architecture that borrows some of the elements of Classical architecture without adhering to the Classical rules.

FREESTONE — stone which can be cut freely in any direction and is relatively easy to tool.

FRIABLE — a term applied to stones that weather easily as the constituent particles loosen and fall away e.g. many soft sandstones break down into loose sand.

GALL — see CLAY GALL.

GLAUCONITE — a mineral occurring as green or green/black grains in sedimentary rocks.

GRANITE — a coarse-grained igneous rock formed as molten material cooled at depth. A dense stone, difficult to cut, but one much used in monumental masonry especially in the nineteenth century because of its range of attractive colouring and the potential for a high polish. Now used mostly as cladding on prestigious buildings and also internally as decorative panels. A good engineering stone.

GRIT, GRITSTONE — arenaceous rock where the particles are angular. A good building stone widely used in the Pennine region. Used for grinding e.g. millstones.

GUTTERS — the vertical grooves cut into a bed of rock during quarrying.

GUTTERSTONES — stones specially shaped to assist drainage in stables etc.

HAMMER-DRESSED — stone shaped in the quarry with a stone axe or a stone hammer which is squared at one end — this gives a rough finish.

HERRINGBONE TOOLING — V-shaped grooving on the face of a stone block.

HOGSBACK, HOG'S BACK — a steep-sided hill that is symmetrical in cross-section.

IGNEOUS — rock formed from the cooling of molten material.

INFILL — the rough rubblestone, mortar etc, thrown into the gap between the facing and backing of a wall.

IONIC — one of the Greek orders of architecture; easily identified by the scrolls at the top of a column.

JADD PICK — a type of pickaxe with two cutting edges on either side of (and at right angles to) the handle; used in quarrying for cutting grooves in a bed of stone prior to the insertion of wedges which are used to lever the stone block from the lower bedding plane.

JOINT — a vertical fracture in a rock at right angles to the bedding planes; in sedimentary rocks the joints form as the sediment dries out.

KEYSTONE — the central wedge-shaped voussoir of an arch. When the arch is 'struck' (i.e. the arch scaffolding is removed) the keystone locks the structure together. Often ornamented.

LABEL — a rectangular-shaped hood-mould of projecting stonework around the top of a window to throw off rainwater.

LABEL STOP — a carved boss at the lower end of the label.

LAMINATION — a very thin layer. Shale is a laminated rock built up of thinly-bedded clays. The term laminating is used to describe the peeling away of the face of a stone due to weathering.

LANCET — a narrow, pointed window characteristic of Early English architecture.

LAVA — a fine-grained, extrusive igneous rock associated with volcanoes and lava flows.

LIMESTONE — a widely varying rock with high calcium carbonate content. Some limestones contain a lot of shell fragments, others are purely chemical precipitations devoid of organic matter. Some are good building stones, e.g. the Bath oolites of Jurassic age. Others are more suited to rubblestone e.g. crystalline Carboniferous limestones and rubbly Silurian limestones.

LIMONITE — a mineral exploited for iron. Tends to produce yellow tints in rocks.

MACHICOLATION — a projecting parapet on the walls and towers of castles with openings at the base through which missiles could be hurled down on the enemy.

MANTLE — the zone of the earth beneath the crust comprising dense material at high temperature and thought to have convection currents which affect movements of blocks of the earth's crust.

MARBLE — strictly a metamorphosed limestone but term also applied to hard limestones that are polished to resemble true marble.

MARL — a calcareous mudstone.

METAMORPHIC — rock changed by heat/pressure related to major earth movements. Rocks may be subjected to high-grade metamorphism and almost completely altered or to low-grade metamorphism where the original characteristics are preserved.

MICA — a shiny material that splits into thin plates.

MICACEOUS — rock with a high proportion of mica. When mica is concentrated in layers the rock is fissile and suitable for paving etc.

MIDDLE POINTED — Decorated Gothic architecture.

MODILLION — a small bracket — one of a series aligned beneath a cornice.

MOULDING — a carved strip of stonework with grooves and projections. Sometimes more ornamental with leaf patterns etc.

MUDSTONE — a hardened clayey rock.

MULLION — a vertical bar of stone separating window lights.

OBSIDIAN — a glassy, black volcanic rock formed by rapid cooling of lava.

OLIVINE — a mineral found in igneous rocks such as dolerite.

OOLITH — a spherical rock particle developed round a nucleus.

OOLITIC LIMESTONE — consists of millions of calcareous ooliths formed around shell fragments.

ORIEL — a projecting window supported on brackets or corbels.

OVERBURDEN — rock, sediment, soil etc. above beds of exploitable rock.

PANGAEA — an ancient super-continent thought to have existed before continental drift caused continents to separate.

PARAPET — a protective wall on top of buildings, on balconies, bridges etc. Used for decorative effect.

PEDIMENT — a triangular gable above the entablature of classical buildings. Smaller versions above doors and windows.

PERPENDICULAR — a late-Gothic style of architecture.

PIER — an independent mass of masonry supporting a beam, lintel or the thrust of an arch.

PILASTER — a flat column built into a wall and projecting slightly.

PINCH-BAR — a device (resembling curved scissors) for lifting a stone block. The grip tightens as the stone is hoisted.

PINNACLE — a small turret-shaped feature on top of church towers and buttresses. Ornamental, but helps a buttress to counter the thrust of a nave vault.

PLATEAU — a flat-topped upland.

POLYCHROMY — many different coloured materials in a single building. Victorian architects were partial to the use of polychromy.

PORPHYRYTIC — a rock texture consisting of large grains embedded in a finer groundmass.

PORTICO — a roofed colonnade at the entrance of a building.

QUARRY SAP — the mineral-laden moisture in freshly quarried stone.

QUARTZ — a hard, resistant mineral found in many rocks.

QUARTZITE — a hard type of sedimentary sandstone rich in quartz: a metamorphosed sandstone — very tough and brittle.

QUOIN — a corner stone — usually a large, dressed stone — often of freestone.

RAINSPLASH EROSION — wearing of stones at the base of a wall by rebounding raindrops.

RANDOM — masonry walling that is not coursed. Stones vary in size and shape.

RANDOM-COURSED — masonry not confined to level courses because large stones break through the horizontal coursing at intervals.

ROCK-FACED — stones that are left rough on the face after some tooling with a wide chisel.

RUBBED — stone which has been finely dressed on the face by abrasion with other stone.

RUBBLESTONE — rough stone, hammer-dressed, of various sizes and shapes incorporated in uncoursed (irregular) or random-coursed walling masonry.

RUSTICATION — sinking of the margins around stone blocks to give the illusion of widely-jointed blocks when fixed in a wall. When only the horizontal edges of each block are sunk, banded rustication features on the wall. Margins may be square sunk or bevelled. With square sunk margins it is possible to save time and effort by adjusting the joints between adjacent stones so that the number of cut margins can be halved.

SANDSTONE — arenaceous rock of great variety of colour and texture but quartz is the common constituent.

SCARP AND VALE TOPOGRAPHY — a landscape comprising alternating escarpments and vales.

SCREE — an accumulation of coarse, angular rock debris.

SEASONING — process whereby quarry sap is released from freshly quarried stone. The stone hardens on exposure to the atmosphere.

SEDIMENTARY — rock derived from sediments worn out of pre-existing rocks.

SHALE — thinly bedded argillaceous rock, very fissile.

SHELF SEA — inshore sea located over continental shelf.

SILICA — a mineral common to many rocks, especially sandstones.

SILICEOUS — rocks dominantly of silica of organic and chemical origin.

SILL — an igneous intrusion which is concordant with bedding planes etc.

SILT — sediment with a texture intermediate between clay and sand.

SILTSTONE — a very fine-grained sandstone.

SLATE — strictly, metamorphosed argillaceous rock but the term is also applied to very thinly bedded, fissile sedimentary rocks which yield 'stone slate'.

SOFFIT — the under-surface of various architectural features. In bridge construction soffit stones form the under-side of an arch.

SPALLING — peeling of the surface of a stone block.

STRATIFICATION — the layering of bedding of rocks.

STRING COURSE, STRING — a narrow, projecting course of stonework on the face of a building.

STROMATOPOROIDS — reef-building marine animals.

SYNCLINE — a downfold of rock strata.

TILE-HANGING — the covering of walls with overlapping rows of tiles.

TILESTONE — very thinly laminated and fissile stone suitable for tiling.

TRANSITIONAL — architectural period between Norman and Early English.

TRANSOM — horizontal stone bar dividing window.

TUDOR-GOTHIC — architectural style characteristic of the Elizabethan age.

TUFA — cellular stone derived from deposition of calcium carbonate when streams emerge from limestone.

TUSCAN — an architectural order; simplified version of the Doric order.

UNDERCUT — the deeply carved portions of mouldings and ornamental sculpture.

VEIN — an intrusion of minerals into a rock mass causing coloured streaks.

VERMICULATION — a type of rustication which is achieved by roughly tooling the face of a stone block to simulate worm tracks.

VOUSSOIR — a wedge-shaped stone, one of a series used to round an arch.

VULCANICITY — volcanic action. Can be violently explosive — volcanoes, or gently effusive — lava flows.

WEATHERING — the process by which rocks are worn away in situ. Weathering agents include wind, rain, frost, temperature changes, plants and animals. Weathering may be chemical and/or mechanical. Climate is crucial — desert regions are very prone to mechanical rock shattering induced by diurnal extremes of temperature. Humid regions are more susceptible to chemical weathering.

Index of Buildings

+ Churches

No.	Church	Grid ref.
1	ALBERBURY, St. Michael	SJ358145
2	BASCHURCH, All Saints	SJ423218
3	BATTLEFIELD, St. Mary Magdalene	SJ513173
4	BICTON, Holy Trinity	SJ448148
5	BOURTON, Holy Trinity	SO597964
6	BRIDGNORTH, St. Leonard	SO717934
7	BRIDGNORTH, St. Mary Magdalene	SO717928
8	BROMFIELD, St. Mary	SO483768
9	BUILDWAS, Abbey	SJ644044
10	BURWARTON, St. Lawrence (now a private residence)	
11	CARDESTON, St. Michael	SJ396124
12	CHETWYND, St. Michael	SJ735214
13	CHIRBURY, St. Michael	SO262985
14	CLAVERLEY, All Saints	SO793935
15	CLEETON-ST-MARY, St. Mary	SO611786
16	CLUN, St. George	SO301806
17	CLUNGUNFORD, St. Cuthbert	SO395787
18	DAWLEY MAGNA, Holy Trinity	SJ687065
19	DIDDLEBURY, St. Peter	SO508854
20	FARLOW, St. Giles	SO639806
21	FORD, St. Michael	SJ413138
22	HIGH ERCALL, St. Michael	SJ595174
23	HOPE BOWDLER, St. Andrew	SO476924
24	KYNNERSLEY, St. Chad	SJ673167
25	LLANDYSILIO, St. Tysilio	SJ268194
26	LLANVAIR WATERDINE, St. Mary	SO240764
27	LUDLOW, St. Laurence	SO512746
28	MAESBROOK, St. John Evangelist	SJ304213
29	MARKET DRAYTON, St. Mary	SJ675341
30	MUNSLOW, St. Michael	SO522877
31	MYNDTOWN, St. John Baptist	SO391896
32	NEWCASTLE, St. John Evangelist	SO254826
33	NEWPORT, St. Nicholas	SJ745193
34	OAKENGATES, St. George	SJ709109
35	POOL QUAY, St. John Evangelist	SJ258118
36	QUATFORD, St. Mary Magdalene	SO738907
37	RATLINGHOPE, St. Margaret	SO403969
38	RHYD-Y-CROESAU, Christ Church	SJ241307
39	RICHARD'S CASTLE, All Saints	SO494707
40	RODINGTON, St. George	SJ589144
41	RUYTON-XI-TOWNS, St. John Baptist	SJ395223
42	SAMBROOK, St. Luke	SJ715245
43	SHIPTON, St. James	SO562918
44	SHREWSBURY, All Saints, Castlefields	SJ499132
45	SHREWSBURY, Holy Cross Abbey	SJ498125
46	SHREWSBURY, Holy Trinity, Belle Vue	SJ496117
47	SHREWSBURY, Holy Trinity, Meole Brace	SJ486105
48	SHREWSBURY, St. Alkmund	SJ493124
49	SHREWSBURY, St. Chad	SJ488125
50	SHREWSBURY, St. George, Frankwell	SJ487130
51	SHREWSBURY, St. Giles	SJ507118
52	SHREWSBURY, St. Mary	SJ493126
53	SIDBURY, Holy Trinity	SO684857
54	SMETHCOTT, St. Michael	SO449994
55	STAPLETON, St. John Baptist	SJ471045
56	STOTTESDON, St. Mary	SO673828
57	WELLINGTON, All Saints	SJ651117
58	WELSH FRANKTON, St. Andrew	SJ364332
59	WESTON, St. Luke	SJ565287
60	WOLLASTON, St. John	SJ329123
61	WROCKWARDINE, St. Peter	SJ624121
62	WROXETER, St. Andrew	SJ564083
63	YOCKLETON, Holy Trinity	SJ395101

Houses

A	ADCOTE MANOR	SJ418195
B	ATTINGHAM	SJ550099
C	CONDOVER HALL	SJ495056
D	THE CITADEL, HAWKSTONE	SJ572286
E	MADELEY COURT	SJ695052
F	MORETON CORBET CASTLE	SJ562232
G	PITCHFORD HALL	SJ527043
H	PRIOR'S LODGE, Much Wenlock	SJ625001
I	RUYTON TOWERS (MANOR)	SJ379228
J	STOKESAY CASTLE	SO436817
K	SWEENEY HALL	SJ295266
L	WOODHOUSE	SJ364288
M	YEATON PEVEREY	SJ443188

S Schools

No.	School	Grid ref.
1	OSWESTRY SCHOOL CHAPEL	SJ285293
2	SHREWSBURY SCHOOL (SHREWSBURY LIBRARY)	SJ494128
3	WATTLESBOROUGH SCHOOL	SJ358117
4	WISTANSTOW SCHOOL	SO432856

X Public Buildings etc, Shrewsbury

No.	Building	Grid ref.
1	LORD HILL'S STATUE	SJ506121
2	MUSIC HALL	SJ491124
3	OLD MARKET HALL	SJ491124
4	SALOP INFIRMARY	SJ494125
5	STATION	SJ495128

● Bridges

1	ATCHAM	SJ541093
2	BRIDGNORTH	SO718930
3	LEE BROCKHURST	SJ549268
4	LLANYBLODWEL	SJ242228
5	MONTFORD	SJ433153
6	PONTFADOC	SJ293211
7	RHYDMEREDITH	SJ252225
8	SHREWSBURY, English Bridge	SJ496124
9	SHREWSBURY, Welsh Bridge	SJ488128
10	WHITTERY	SO272983

dex maps (numbers relate to the buildings and bridges listed in the index)

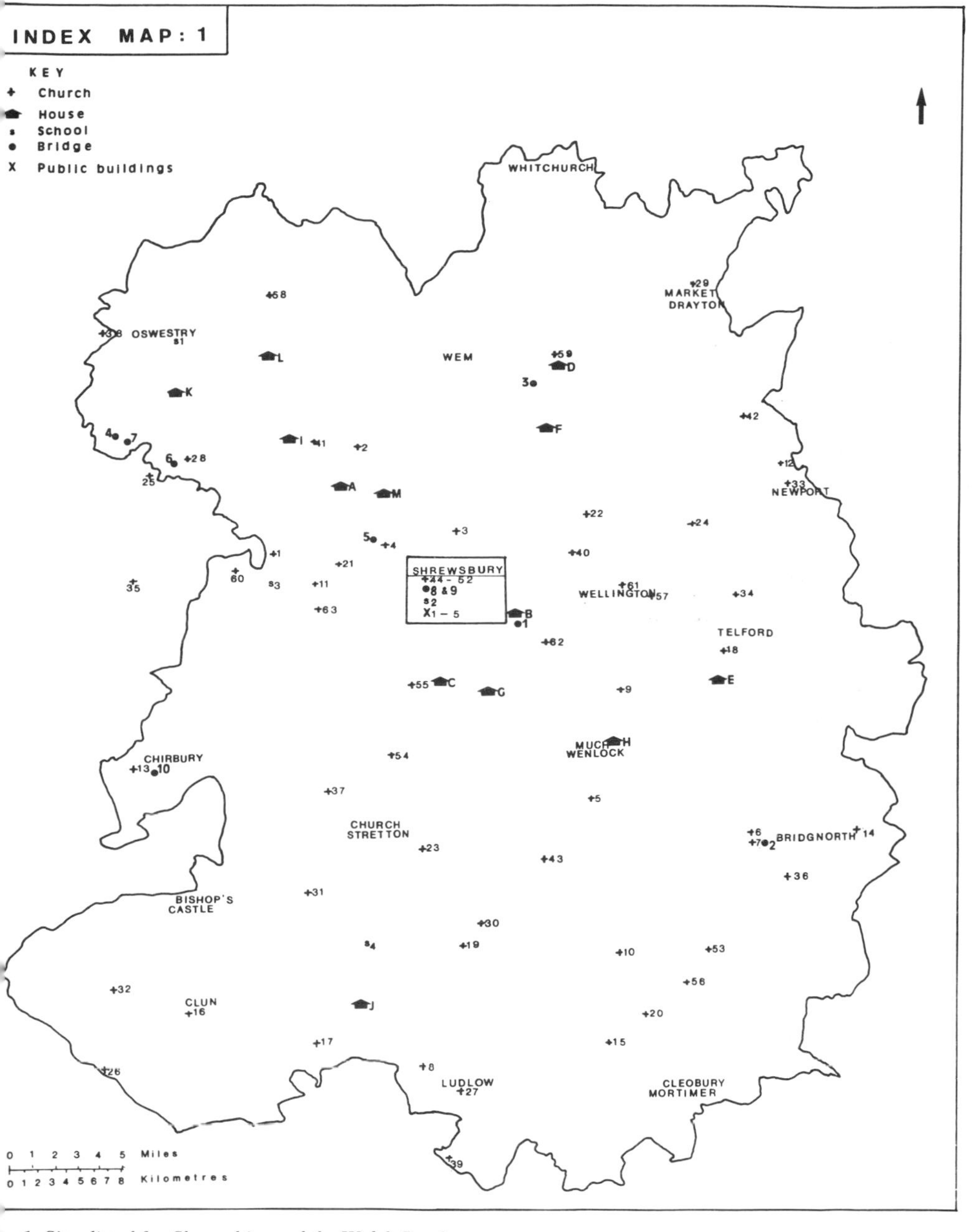

p 1: Sites listed for Shropshire and the Welsh Border

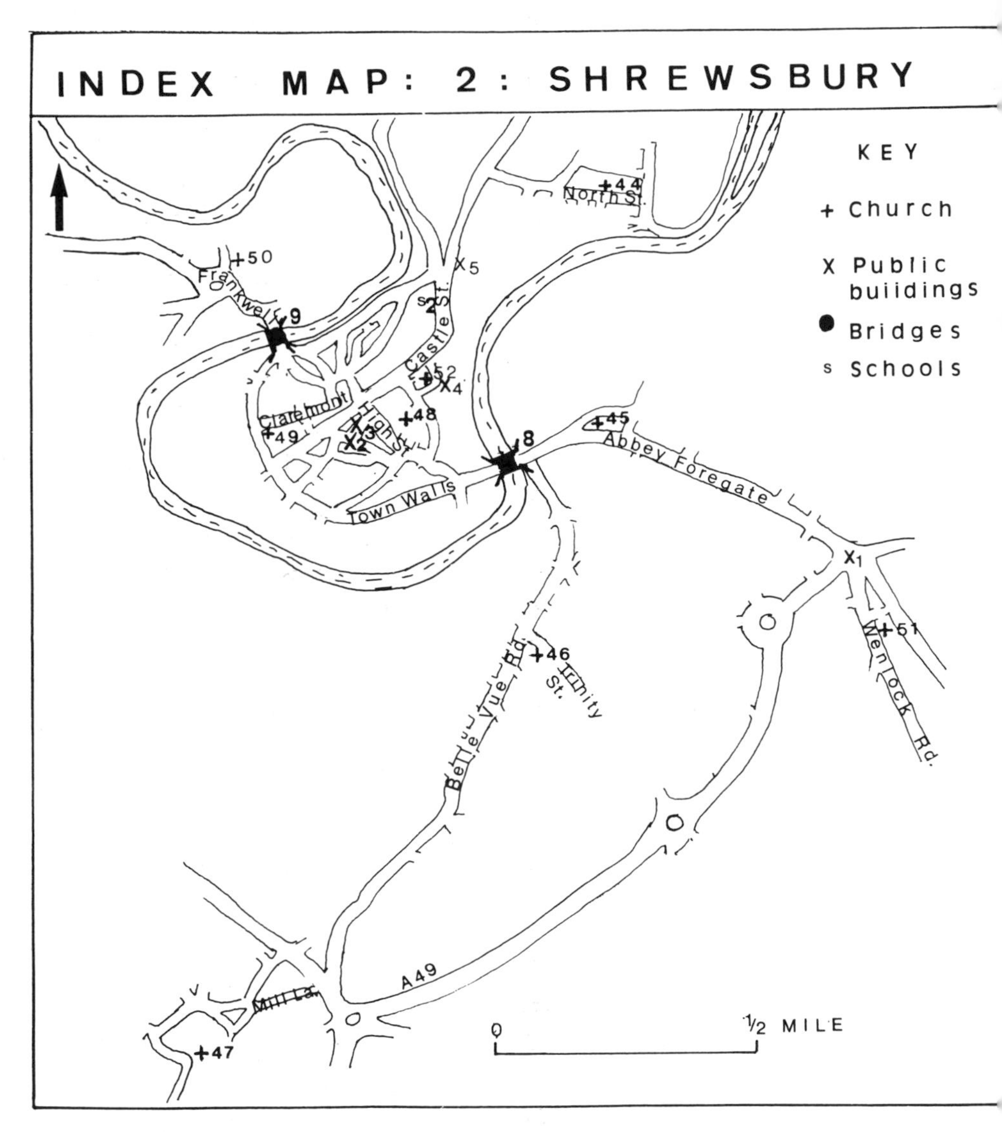

Map 2: Sites listed for Shrewsbury

Bibliography

ARCHITECTURE AND BUILDING MATERIALS — General

Allsop, Bruce, *The Country Life Companion to Architecture in Britain and Europe*, Hamlyn, (Feltham) 1985

Ashurst, John and Dimes, Francis G., *Stone in Building — its Use and Potential Today*, Architectural Press, (London) 1977

Barman, Christian, *An Introduction to Railway Architecture*, Art and Technics, (London) 1950

Binney, Marcus and Pearce, David, *Railway Architecture*, Bloomsbury Books, (London) 1985

Braun, Hugh, *Parish Churches, Their Architectural Development in England*, Faber and Faber, (London) 1970, second edition 1974

Braun, Hugh, *A Short History of English Architecture*, Faber and Faber, (London) 1956, second edition 1978

Briggs, Martin S., *Everyman's Concise Encyclopaedia of Architecture*, Dent, (London) 1959, reprinted 1978

Brunskill, R.W., *Illustrated Handbook of Vernacular Architecture*, Faber and Faber, (London) 1971, 1978

Clifton-Taylor, Alec, *English Parish Churches as Works of Art*, Batsford (London) 1974

Clifton-Taylor, Alec, *The Pattern of English Building*, Faber and Faber, (London) 1972

Clifton-Taylor, Alec and Ireson, A.S., *English Stone Building*, Victor Gollancz, (London) 1983

Davey, N., *Building Stones of England and Wales*, Bedford Square Press, (London) 1976

Davey, N., *A History of Building Materials*, Phoenix House (London) 1961

Dixon, Roger and Muthesius, Stefan, *Victorian Architecture*, Thames and Hudson, (London) 1978

Girouard, Mark, *The Victorian Country House*, Yale University Press, (Newhaven and London) 1979

Harris, Robert, *Canals and Their Architecture*, Godfrey Cave Associates, (London) 1980 (first published by Hugh Evelyn Ltd., 1969)

Hudson, Kenneth, *Building Materials*, Longman, (London) 1972

Jordan, Robert Furneaux, *Victorian Architecture*, Penguin, (Hardmondsworth) 1966

Meeks, Carroll L.V., *The Railway Station, An Architectural History*, The Architectural Press, (London) 1957

Mercer, Eric, *English Vernacular Houses*, HMSO, (London) 1975

O'Neill, Hugh, *Stone for Building*, Heinemann, (London) 1965

Saint, Andrew, *Richard Norman Shaw*, Yale University Press, (Newhaven and London) 1976

Sealcy, Antony, *Bridges and Aqueducts*, Hugh Evelyn Ltd., (London) 1976

Shore, B.C.G., *Stones of Britain, A Pictorial Guide to Those in Charge of Valuable Buildings*, Leonard Hill Ltd., (London) 1957

Smith, Edwin and Cook, Olive and Hutton, Graham, *English Parish Churches*, Thames and Hudson, (London) 1976

LOCAL ARCHITECTURE AND LANDSCAPE HISTORY

Blackwall, Anthony, *Historic Bridges of Shropshire*, Shropshire Libraries, (Shrewsbury) 1985

Burke's and Savills Guide to Country Houses, Vol. 2, Burke's Peerage, Ltd., (London) 1980

Carr, A.M., *Shrewsbury Library, its History and Restoration*, Shropshire Libraries, (Shrewsbury) 1983

Clifton-Taylor, Alec, *Six English Towns*, BBC, (London) 1978

Cranage, D.H.S., *An Architectural Account of the Churches of Shropshire*, Hobson and Co., (Wellington, Shropshire) 1903

Forrest, H.E., *Some Old Shropshire Houses*, published by author, (Shrewsbury) 1924

Haslam, Richard, *The Buildings of Wales, Powys*, Penguin Books, (Harmondsworth) 1979

Hill, Mary, C., *The History of Shropshire's many Shirehalls*, Shropshire County Council, (Shrewsbury) 1963

Jenkins, A.E., *Titterstone Clee Hills, Everyday Life, Industrial History and Dialect*, published by author, (Ludlow) 1982

Leighton, Stanley, *Shropshire Houses Past and Present*, George Bell and Sons, (London) 1901

Millward, Roy and Robinson, Adrian, *The Welsh Borders*, Eyre Methuen, (London) 1978

Millward, Roy and Robinson, Adrian, *The Welsh Marches*, Macmillan, (London) 1971

Moulder, Michael, *Shropshire, A Shell Guide*, Faber and Faber (London) 1972

Pevsner, Nikolaus, *The Buildings of England, Shropshire*, Penguin, (Harmondsworth) 1958, reprinted 1979

Pidgeon, Henry, *Memorials of Shrewsbury*, Salop County Library, (Shrewsbury) 1837

Rowley, Trevor, *The Landscape of the Welsh Marches*, Michael Joseph, (London) 1986

Rowley, Trevor, *The Shropshire Landscape*, Hodder and Stoughton, (London) 1972

Trinder, Barrie, *A History of Shropshire*, Phillimore, (Chichester) 1983

Victoria County History of Shropshire, The University of London Institute of Historical Research

Ward, Arthur Walburgh, *The Bridges of Shrewsbury*, Wilding and Son Ltd., (Shrewsbury) 1935, reprinted by Shropshire Libraries, (Shrewsbury) 1983

Woolley, George et al., *The Parish Church of St. Laurence, Ludlow. A Monograph of the Restoration, 1889-90-91*, George Woolley, (Ludlow) 1893

GEOLOGY AND LANDSCAPE

N.B. Geological Survey Memoirs and Sheet Explanations are listed below the relevant maps.

Dean, W.T., *Geological Itineraries in South Shropshire*, Geologists' Association, Guides No. 27, Benham and Company, (Colchester) 1968

Earp, J.R. and Hains, B.A., *British Regional Geology, The Welsh Borderland*, HMSO, (London) third edition 1971

Hains, B.A., and Horton, A., *British Regional Geology, Central England*, HMSO, (London) third edition 1969

Toghill, Peter and Chell, Keith, *Shropshire Geology, Stratigraphic and Tectonic History*, Field Studies Council 163, Publication G23, 1984

Trueman, A.E., *Geology and Scenery in England and Wales*, Penguin, (Harmondsworth) 1963

MAPS

ORDNANCE SURVEY 1 : 50,000

Sheet 117 'Chester'
Sheet 125 'Bala and Lake Vyrnwy'
Sheet 126 'Shrewsbury'
Sheet 127 'Stafford and Telford'
Sheet 137 'Ludlow and Wenlock Edge'
Sheet 138 'Kidderminster and the Wyre Forest'

ORDNANCE SURVEY 1 : 25,000 First Series

Sheet SJ 22 'Oswestry (South)'
Sheet SJ 23 'Chirk'
Sheet SJ 24 'Llangollen'
Sheet SO 38 'Bishop's Castle'
Sheet SJ 41 'Shrewsbury (West)'
Sheet SO 48 'Craven Arms'
Sheet SO 49 'Church Stretton'
Sheet SO 59 'Wenlock Edge (North)'
Sheet SJ 60 'Ironbridge'
Sheet SJ 61 'Wellington'
Sheet SJ 70 'Shifnal'

ORDNANCE SURVEY 1 : 25,000 Pathfinder Series

Sheet SO 29/39 'Montgomery'
Sheet SJ 20/30 'Welshpool (Trallwng)'
Sheet SJ 21/31 'Middletown and Nesscliffe'
Sheet SJ 22/32 'Oswestry (Croesoswallt)'
Sheet SJ 23/33 'Chirk (Y Waun)'
Sheet SJ 24/34 'Llangollen and Wrexham (Wrecsam) South'
Sheet SJ 25/35 'Wrexham (Wrecsam) North'
Sheet SJ 40/50 'Dorrington and Cressage'
Sheet SJ 41/51 'Shrewsbury'
Sheet SJ 42/52 'Wem and Myddle'
Sheet SJ 43/53 'Ellesmere (East) and Prees'
Sheet SO 47/57 'Ludlow'
Sheet SO 48/58 'Craven Arms'
Sheet SJ 62/72 'Hodnet and Norbury'
Sheet SJ 63/73 'Market Drayton'
Sheet SO 67/77 'Wyre Forest and Cleobury Mortimer'
Sheet SO 68/78 'Highley'
Sheet SO 69/79 'Bridgnorth and Much Wenlock'

GEOLOGICAL SURVEY 1 : 63,360 or 1 : 50,000 New Series

Sheet 121 'Wrexham' Solid
Sheet 122 'Nantwich' Solid
Sheet 137 'Oswestry' Solid
Sheet 138 'Wem' Solid and Drift
Sheet 139 'Stafford' Drift
Sheet 152 'Shrewsbury' Solid
Sheet 153 'Wolverhampton' Solid
Sheet 166 'Church Stretton' Solid
Sheet 167 'Dudley' Solid and Drift

GEOLOGICAL SURVEY 1 : 25,000 (about 2½ inches to 1 mile)

Sheet SO 48 'Craven Arms' Solid and Drift
Sheet SO 49 'Church Stretton' Solid and Drift
Sheet SO 59 'Wenlock Edge' Solid and Drift
Special Sheets, Parts of SO 47, 57, 'Leintwardine-Ludlow' Solid

GEOLOGICAL SURVEY MEMOIRS AND SHEET EXPLANATIONS

Greig, D.C., Wright, J.E., Hains, B.A. and Mitchell, G.H., *Geology of the Country around Church Stretton, Craven Arms, Wenlock Edge and Brown Clee*, Explanation of Sheet 166 (1968)

Hains, B.A., *The Geology of the Craven Arms area*, Explanation of Sheet SO 48 (1969)

Hains, B.A., *The Geology of the Wenlock Edge area*, Explanation of Sheet SO 59 (1970)

Pocock, R.W., Whitehead, T.H., Wedd, C.B., and Robertson, T., *Shrewsbury District including the Hanwood Coalfield*, Explanation of Sheet 152 (1938)

Pocock, R.W. and Wray, D.A., *The Geology of the Country Around Wem*, Explanation of Sheet 138 (1925)

Wedd, C.B., Smith, B., King, W.B.R. and Wray, D.A., *The Country Around Oswestry*, Explanation of Sheet 137 (1929)

Whitehead, T.H., Dixon, E.E.L., Pocock, R.W., Robertson, T. and Cantrill, T.C., *The Country between Stafford and Market Drayton*, Explanation of Sheet 139 (1927)

Whitehead, T.H. and Pocock, R.W., *Dudley and Bridgnorth*, Explanation of Sheet 167 (1947)

Whitehead. T.H., Robertson, T., Pocock, R.W., and Dixon, E.E.L., *The Country between Wolverhampton and Oakengates*, Explanation of Sheet 153 (1928)

Wright, J.E., *The Geology of the Church Stretton area*, Explanation of Sheet SO 49 (1968)